SHaKe UP

SCIENCE 6

Pearson Education Limited
Edinburgh Gate
Harlow
Essex CM20 2JE
England
and Associated Companies throughout the world.

www.pearsonelt.com

First published 2016

ISBN: 978-1-2921-4484-9

Set in Futura LT Pro, Feltpen Com, Bauhaus Std, ITC Benguiat Gothic Std, ITC Zapf Dingbats Std

Printed in China (CTPSC/01)

Acknowledgments
Picture Credits

The publisher would like to thank the following for their kind permission to reproduce their photographs:

(Key: b-bottom; c-centre; l-left; r-right; t-top)

123RF.com: 16tl, 16tc, 16tr, 19tr, 40tl, 40tc (left), 42cr, 43b (l), 48bl, 58t, 58b, 63b, 64tc (left), 64tr, 67t, 71, 75b, 76b, 90b, 93b, 97tl, 97b, 100tr, 101t, 102tl, 106tc, 109tr, 111br, alhovik 66bl, alzam 68c, aquafun 17cr, Gary Arbach 5tl, arinahabich 20t, Karina Baumgart 40cr, Kristyna Bazantova 18br, Roberto Biasini 32r, blueringmedia 30cl, Linda Bucklin 19b, 22b, 27tl, George Burba 54br, 56, Ken Byrne 17b, Leonello Calvetti 28tc, 28tr, 31, 34tl, Ann Cantelow 54cr, choreograph 81b, Narongrit Dantragoon 46tl, 46tc, dcylai 97tc, 99bc, Melinda Fawver 45t, Adam Gryko 48br, hadot 49c, Hafakot 4r, Sebastian Kaulitzki 29, Galina Khaydukova 40cl, David Koscheck 59c, Anna Kukhmar 109tl, 111bl, lightwise 8t, Jim Mills 69tl, Christian Mokri 16b, Wuttichok Panichiwarapun 46tr, Ghenadie Pascari 10, Petro Perutskyi 59b, pictureguy66 46b, ponsuwan 15b, Steven Prorak 49bc, Przemyslaw Przybylski 7bl, Oleksandr Rybitskyi 48t, Pablo Scapinachis Armstrong 82t, Fedor Selivanov 20b, Michael Shake 48bc, 51br, Dmitriy Shironosov 24b, Nico Smit 60b, Janis Smits 48c, Martin Spurny 49tc, strannikfox 48l, tempusfugit 82b, Narith Thongphasuk 44br, tlorna 35bl, Tom Wang 96b, Cathy Yeulet 88tr; **Alamy Images:** moodboard 23bl, National Geographic Image Collection 4bl, RGB Ventures 13t, Stocktrek Images, Inc. 69b, Sam Toren 7bc, Westend61 GmbH 100bl; **Fotolia.com:** Giordano Aita 64tc (right), anankkml 76cr, berzina 52b, BillionPhotos.com 12t, bluesnote 84t, 103br, Chepko Danil 40b, destina 65b, Elenathewise 76tc (left), 107tc, Gajus 47t, Graphies.thèque 54l, H Grose 106bl, herraez 22t, highwaystarz 9, illustrez-vous 96t, Budimir Jevtic 99cr, johnnydao 13b, kevron2001 65t, korsaralex 68b, maxsol7 76tr, 82c, mercava2007 85t, 87b, Mezzotint 6t, mouse_md 50, Nightman1965 61, OConnor 41, Igor V. Podkopaev 43t, samsonovs 12b, scottchan 78b, SergiyN 11b, Dmitriy Sladkov 78t, Snaptitude 91, storm 65c, Natasa Tatarin 45b, James Threw 66t, tmass 72b, Vivid Pixels 39bl, Tom Wang 53t, zlikovec 84c, zurbagan 64tl; **Getty Images:** James Balog 15t, John Cancalosi 23br, 27b, Andrew Geiger 6b, Kansas City Star 72t, Photo-Biotic Leigh Righton 78tc; **Glow Images:** Image 100. Corbis 88tc; **Imagestate Media:** John Foxx Collection 17tl, 27tr, 57, 88tl; **Pearson Education Ltd:** Martyn F Chillmaid 52tc, Mohd Suhail. Pearson India Education Services Pvt. Ltd 37t, Oxford Designers & Illustrators Ltd 30b, 77t, 108b; **SF Glenview Photo Studio:** 14, 26, 38, 62, 74, 86, 98, 110; **Shutterstock.com:** 501room 25tc, 21r, 40tc, 44bl, 106c, Anita Patterson Peppers 103bl, ArTDi101 40tc (right), Benjamin Marin Rubio 104b, Stephane Bidouze 49b, Barry Blackburn 41b, BlueRingMedia 28tl, 36tr, boban_nz 15c, Natalia Bratslavsky 52tr, Eric Broder Van Dyke 79t, bunnyphoto 102cr, C Eng-Wong Photography 49t, Leonello Calvetti 33, Jose Luis Calvo 34tr, Cessna152 93t, chaoss 102b, cobalt88 77br, Kevin Connors 99t, Danshutter 102tc, Sam DCruz 64cr, Andrea De Paoli 83tr, defpicture 100tl, Willy Deganello 97tc (left), dencg 47c, Chris DeRidder 76cl, design56 92, Designua 53b, 110r, djgis 85br, EBFoto 106cl, Shilova Ekaterina 8br, ene 106tl, EpicStockMedia 90, Paul Fleet 67b, 75t, fotoJoost 63cl, Gentoo Multimedia Limited 101b, glo 94, gregdx 89b, Karen Grigoryan 97cr (bottom), 99bl, Martin Haas 88b, Happy Together 88cr, Jeff Hinds 102tr, Holbox 17tr, Lorraine Hudgins 25bc (right), Ivancovlad 88cl, Sarah Jessup 90c, Monica Johansen 25b, kangshutters 108tr, Raymond Kasprzak 19tl, Kichigin 40c, 51bl, Mikhail Kolesnikov 68t, Georgios Kollidas 107tr, Olga Kovalenko 11t, Ivan Kuzmin 106br, Kzenon 39br, kzww 97cr (top), Lepas 76tc (right), Loskutnikov 109b, majeczka 58bc, Marko Marcello 80b, Marian.P 63c, matin 85bl, Joze Maucec 18bc, Andrew McDonough 95b, 99br, Michelle D. Milliman 100br, Minton 58tc, Alex Mit 34b, Mopic 73t, 109tc, 111bc, MrGarry 108tc, Andrii Muzyka 36tl, naluwan 104t, Maksim Nikalayenka 64b, Ociacia 5r, Olgysha 103t, ollyy 106cr, Tyler Olson 60t, Palto 95t, A Paterson 25bc (top left), Pedrosala 107br, Mike Phillips 51t, Phloxii 35br, Photobar 25t, photokup 37b, Paisarn Praha 17tc, 23t, Belinda Pretorius 52tl, Lee Prince 54tr, 63t, psamtik 25bc (bottom left), Pylypenko 24t, Cheryl Ann Quigley 89t, R. Mackay Photography 69tr, Joshua Resnick 76tl, 81t, Laurin Rinder 70, Julija Sapic 97tc (right), Ilin Sergey 8bc, Mihai Simonia 102cl, Ben Smith 28b, Ivan Smuk 108tl, Audrey Snider-Bell 18bl, Somchai Som 97tr, ssuaphotos 93c, Stana 40tr, stockshoppe 100c, Stocksnapper 88c, StudioSmart 107tl, Studiotouch 21l, 27tc, swa182 47b, Swapan Photography 59t, 63cr, Sychugina 87t, Artur Synenko 42t, Dan Tautan 51bc, Max Topchii 79c, Tomasz Trojanowski 32t, Jeroen van den Broek 8c, Wallenrock 76c, Shi Yali 55, Yellowj 80t, Feng Yu 106tr, Sergiy Zavgorodny 83cl, Arman Zhenikeyev 35t All other images © Pearson Education

Every effort has been made to trace the copyright holders, and we apologize in advance for any unintentional omissions We would be pleased to insert the appropriate acknowledgment in any subsequent edition of this publication.

Science Consultant
Mark Sander

Cover images: *Front:* **Getty Images:** laflor r; **Shutterstock.com:** Ociacia l; *Back:* **Shutterstock.com:** 501room l, Belinda Pretorius c, ssuaphotos r

Contents

Unit 1 — Design and Function

How can technology make our lives easier?

1 With a partner, make a list of recent inventions and new technology.

2 With a partner, number the first five steps of the design process.

- ☐ Do research.
- ☐ Design and construct a prototype.
- ☐ Identify the problem.
- ☐ Choose one solution.
- ☐ Develop possible solutions.

3 What problems were these robots invented to solve? Discuss with a partner.

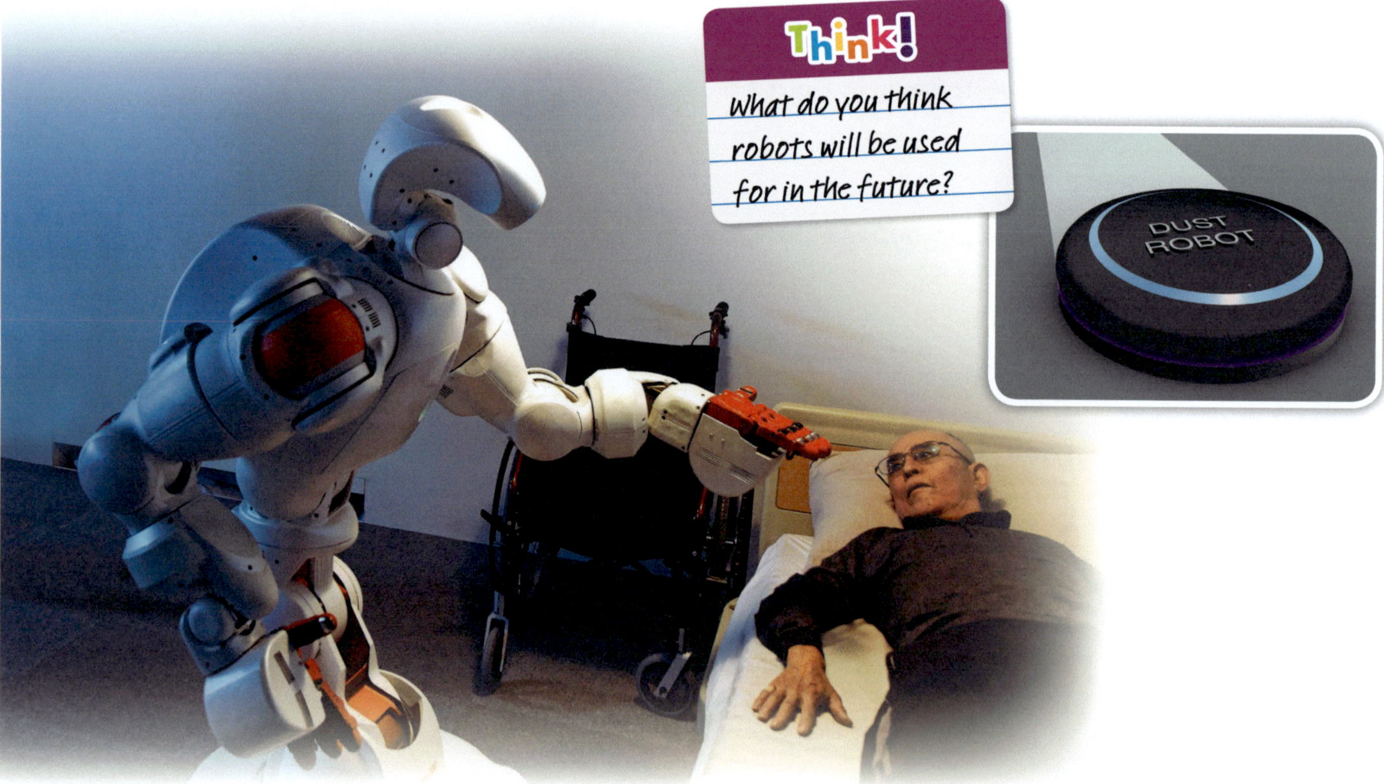

Lesson 1 · How does technology mimic living things?

1 How can this device help someone communicate? Discuss as a class.

2 Read and circle *T* (true) or *F* (false).

Technology and the Human Body

The human body is an amazing structure. Engineers sometimes use scientific knowledge of how the body works to develop technologies. Some of the technologies help people whose bodies do not function as they should. Some technologies do tasks that are too dangerous for people. Technologies that have moving parts can be like the human body. A robot is one of these technologies. Robots can have a body structure and movable joints that are similar to the human skeletal and muscular systems. Robots use an electrical energy source to help them move. The human body uses energy from food to help it move. Robots have a **sensor system** and a computer to control movement. In the human body, the brain and nervous system help to control movement.

1. Technology can benefit people with disabilities.   T / F

2. Robots can mimic parts of the human body.   T / F

3. Robots have nervous systems to control movement.   T / F

4. Robots can use a variety of energy sources to operate.   T / F

3 In what ways do you think a robot is like you? Discuss with a partner.

4 How is the runner's prosthetic leg similar to the legs of the other runners? Discuss as a class.

5 Read and, with a partner, discuss the questions below.

Prosthetic Limbs

Robotic technology can also be used to make a prosthetic limb move. A **prosthetic limb** is an artificial arm, hand, leg, or foot that replaces a missing one. Modern prosthetic limbs can be controlled by electrical signals from the brain.

In the past, prosthetic hands had few fingers and could not do many things. Today, they have a thumb and four fingers that are controlled individually. These prosthetic hands can turn a key, pick up small objects, and hold a glass.

Current prosthetic legs and feet allow their users to walk and even run. As technology advances, prosthetic legs and feet work more like real legs and feet. The latest prosthetic limbs also look more like real limbs.

1. How do prosthetic limbs help people?
2. How are prosthetic hands different today compared with those of the past?

6 What does this man's prosthetic hand enable him to do? Look and discuss with a partner.

  ▷ **I Will Know...**

7 **Read. With a partner, list the characteristics of each machine or robot that mimics the real animal.**

Animals and Technology

Some technologies mimic the muscular and skeletal systems of animals. These systems help animals to move in different ways. The wings and tails of birds help them fly. Fish have muscular and skeletal systems that help them swim.

Airplanes have parts that mimic the wings and tails of birds. Like the wings and tails of birds, airplane wings and tails can be adjusted to control how the airplane moves.

Some robots can also fly. The robotic bat flaps its wings and flies like a bat. It can search collapsed buildings and other areas people cannot get to. Some robots that are used to explore the ocean have parts that mimic the muscular and skeletal systems of fish.

Scientists use robotic animals to study the behavior of real animals. A robotic squirrel makes noise and moves its tail like a real squirrel. It can be placed in an area where real squirrels live. A real squirrel may wiggle its tail and make noises at the robotic squirrel. Scientists can use this information to learn how squirrels communicate with one another.

airplane/bird	robotic bat/bat	robotic squirrel/squirrel

8 **As a class, discuss how animal robots can be used.**

real squirrel

robotic fish

At-Home Lab

Walk around your neighborhood with an adult. Observe any ways in which technology mimics living things. Record these observations in your Science Notebook.

9 **Read and circle *T* (true) or *F* (false).**

Nanobots

How can you build a robot that is only a few billionths of a meter long? Scientists hope to be able to build these tiny robots using **nanotechnology**. Scientists have found ways to move one **atom** at a time. They hope to be able to use this technology to build tiny robots, or **nanobots**, that can perform all kinds of tasks.

One idea is to use nanobots inside the human body. Nanobots may be able to deliver medications better than current methods. Scientists are also researching how to make a nanobot that can remove cholesterol from the walls of arteries.

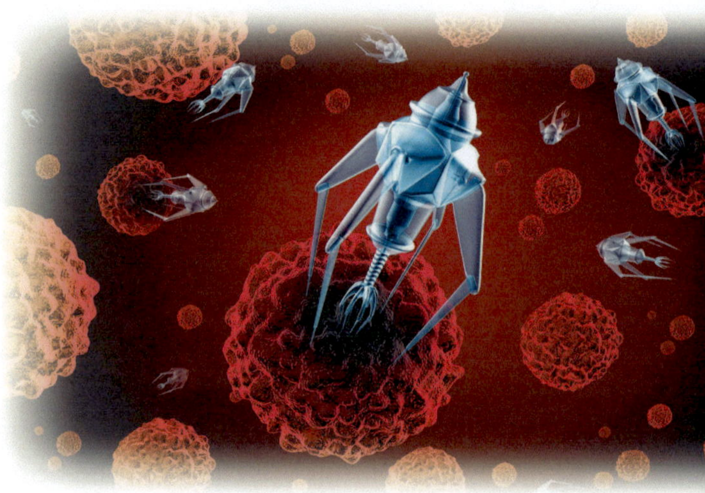
Scientists hope that nanobots will be able to kill cancer cells or treat other human disease.

1. Tiny robots called nanobots already exist.   **T / F**

2. Scientists believe nanobots will not be used inside human bodies.   **T / F**

3. Nanobots may be able to help sick people more effectively than current methods.   **T / F**

10 **Invent a technology that could help a blind person. Draw your prototype and present your idea to the class.**

In what other ways do you think nanotechnology might be useful?

This nanotube's walls are only one atom thick!

Lesson 2 · What is the design process?

1 **Read and underline the word for a person who designs technology.**

Design Process

Technology helps to solve many of the problems we have. We use technology in our homes, schools, and offices. There are technologies for constructing buildings, communicating with others, transporting people and products, and so much more.

Who makes all this technology? People all over the world develop technologies. You may be surprised to know that even students your age develop new technologies. An engineer is a person who designs new technologies. People work in many different fields to apply scientific knowledge to everyday life. People use the **design process** to develop new technologies. The design process is a set of steps for developing products and processes that solve problems.

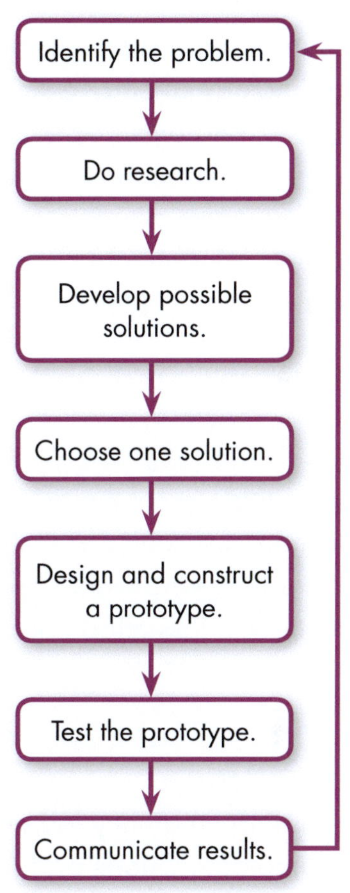

2 **Why do you think it is important to follow the design process when designing new technologies? Discuss with a partner.**

3 **Look at the steps of the design process. In your notebook, write what you think each step involves.**

4 **Read and write the heading for each stage of the design process.**

> Identify the Problem   Develop Possible Solutions   Do Research

---

In order to make or improve existing technology, scientists need to know what technology already exists. Scientific journals, magazines, the Internet, informational books, and encyclopedias can be helpful for solving design problems. Interviewing an expert may be the best way to find out information.

Engineers designing a new toothbrush might investigate how the shape of the handle affects how people brush their teeth. Engineers should also know how different **bristle** materials affect teeth.

In this step of the design process, it is necessary to identify a need or problem. All technology comes from the need for a solution to a problem. It is important in this step to determine who would be helped by the solution. For example, a toothbrush that cleans teeth with less effort could potentially help everyone reduce cavities and gum problems.

Using what they learned, scientists and engineers think of ways to improve an existing technology. Charts and diagrams can be useful to communicate their design solutions.

5 **Underline the sources you would use to find out what other scientists are working on. Then circle how scientists can communicate their design solutions.**

6 **Look and, with a partner, explain what you can learn from the drawing and what step of the design process it corresponds to.**

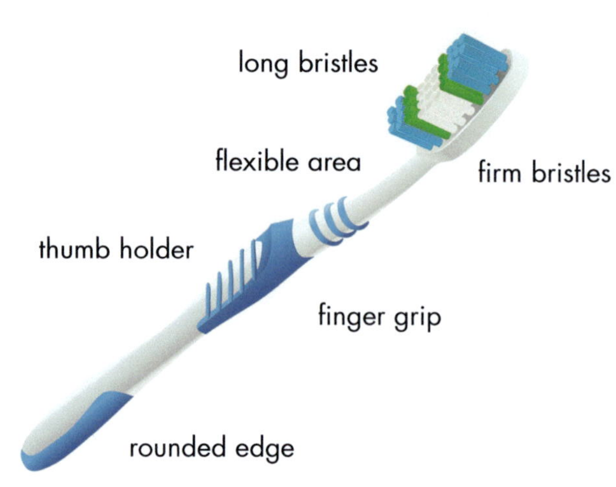

7 **Read and answer the questions below.**

Choose one solution.

It is important to choose wisely the one solution you will build. Making many solutions may take too much time. The cost of making the solution can also affect your decision. For example, even if the toothbrush works very well, people may not buy it if it is very expensive.

Design and construct a prototype.

The next step is to build a model of the solution, called a **prototype**. It is used to test the solution. It is important to identify the kinds of materials you use to build your prototype. The properties of the materials you use affect the function of your prototype. You will need a strong, flexible material for parts that bend. If you do not want the part to bend, you should use a rigid material. You will also need to identify the tools you use to build your prototype.

toothbrush prototype

Test the prototype.

The prototype needs to be tested to see if it meets the requirements to solve the problem. Engineers make careful measurements as they test their prototypes. When testing a toothbrush, engineers might measure how much **plaque** is left on the teeth after brushing for one minute. These measurements help the engineers evaluate how well the prototype works.

1. What are two things to consider when building a prototype?

   a) _______________________________   b) _______________________________

2. Why do engineers build prototypes of their design solutions? Circle.

   a) To test it to ensure that it works.   c) To test the most expensive model.
   b) To present it to others as a final solution.

3. Read and fill in the blank.

   When testing a prototype, engineers must take careful _______________________ to ensure they are evaluating it accurately.

8 Read and underline two reasons it is important to communicate your results to others. Then, in pairs, fill in the chart.

Communicate Results

Throughout the design process it is important to document your work. **Document** means to record what you learn. Documentation helps you communicate with others. If you are working in a company, you will need to communicate your process and design to managers, salespeople, and many others. Often others will need to repeat your tests to verify the results. They will need to know your test procedures and the specifics of your design. The people you share your design with may be able to offer advice on how to improve your idea.

Your design solution can be communicated in many ways. Labeled diagrams can show the size and shape of the parts of your product. Graphic organizers can show how the parts are put together. You will also need a list of materials and tools used to make each part. Tables, charts, and graphs can help you communicate test results.

Ways to Communicate Your Design Solution	
1. _______________________	4. _______________________
2. _______________________	5. _______________________
3. _______________________	6. _______________________

9 Read and look. Then, in pairs, make a list of three ways the redesigned prototype is better than the original.

Evaluate and Redesign

Using the results of your tests and feedback from others, you can evaluate how well your design solved the problem. This information can help you redesign your product to make it work better. You may need to make minor adjustments or choose a completely new solution.

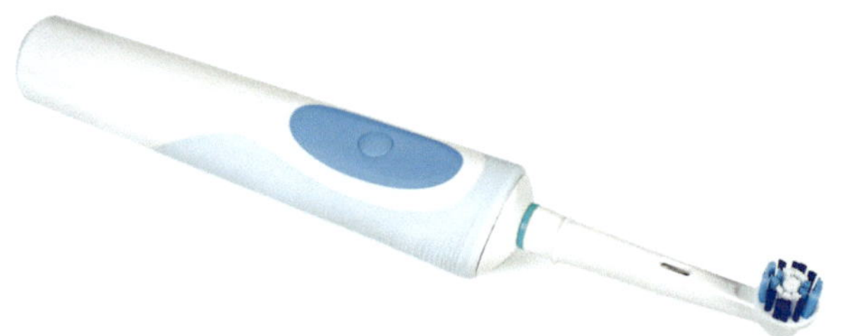

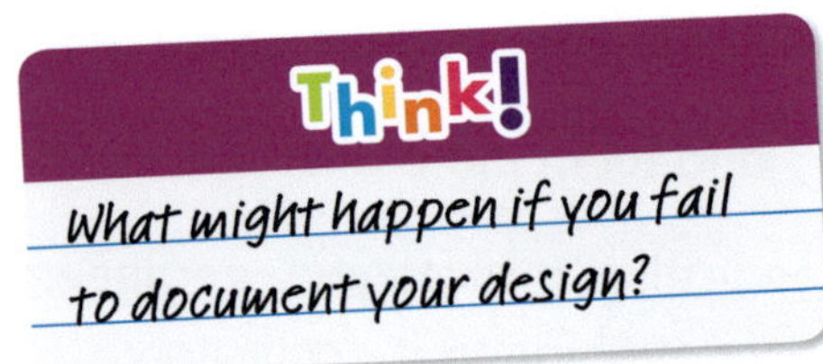

10 **Read and complete the answers. Then match the answers with their questions.**

Designing Robotic Arms

Engineers use the design process to develop robotic arms. **Robotic** arms are designed and built to mimic the movement of human arms.

The first robotic arm used in a factory was developed by George Devol. The robotic arm picked up and stacked metal parts that were too hot for workers to handle. George Devol and his partner, Joseph Engelberger, called the robotic arm the *Unimate*.

The *Unimate* had a "shoulder" but no "elbow." Devol and Engelberger continued to redesign the robotic arm. They developed a new robotic arm with an "elbow" that allowed it to perform more tasks. Today's robotic arms can move in many different directions.

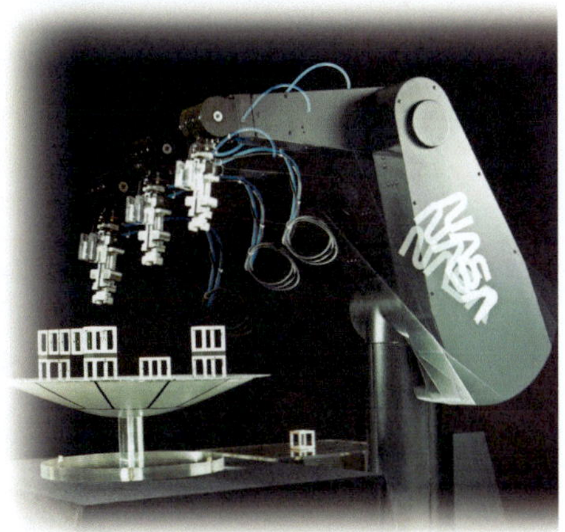

PUMA, *an industrial programmable robot, was introduced in 1980.*

1. Who designed the first robotic arm?

2. What could the first robotic arm do?

3. What did the *Unimate* not have?

The *Unimate* didn't have an
___________________________.

____
developed the first robotic arm.

The first robotic arm was able to
____
___________________________.

11 **Look and label the parts of this modern robotic arm that are like a human arm.**

shoulder   wrist   elbow

12 **With a partner, discuss why the modern robotic arm might be better than *PUMA*.**

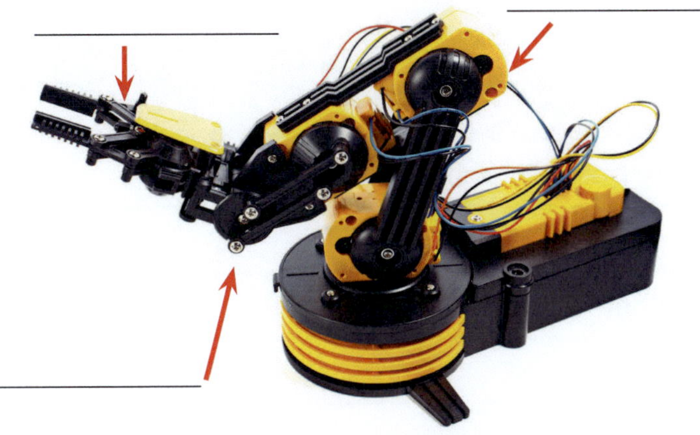

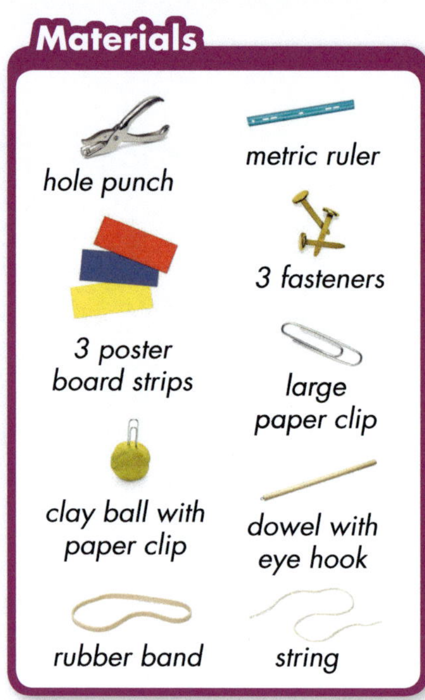

Let's Investigate!

How can you make and redesign a model of a robotic arm?

1. Use a hole punch to make holes in three poster board strips as shown. Use two fasteners to join the strips together.

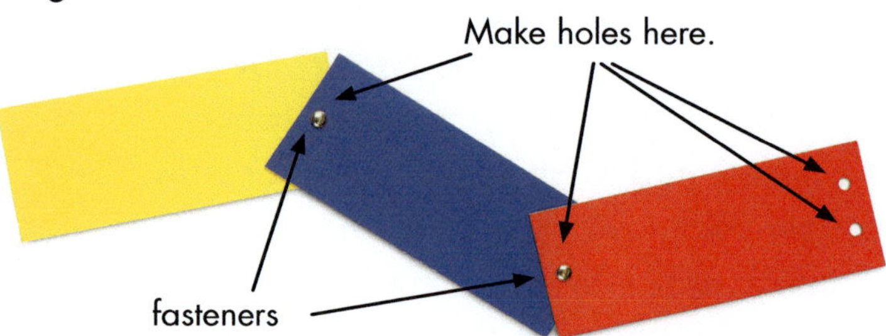

2. Use a fastener. Attach the eye hook on the dowel to one of the two holes on the red strip.

3. Bend a large paper clip into an S shape and put the top of the S through the other hole in the red strip.

4. Use the robotic arm. Try to pick up the objects listed in the chart. Record the number of tries you need. Use up to 5 tries for each object.

5. Redesign your model of a robotic arm. Repeat step 4.

Objects Chart		
Object	**Number of Tries**	
Clay ball with paper clip		
Paper clip		
Rubber band		
String		

REVIEW THE BIG Q

How can technology make our lives easier?

Lesson 1

How does technology mimic living things?

1 **Circle the correct answer.**
An artificial arm that mimics the real human muscular and skeletal system is called a _________________ arm.

a) synthetic
b) prosthetic
c) limb
d) mechanical

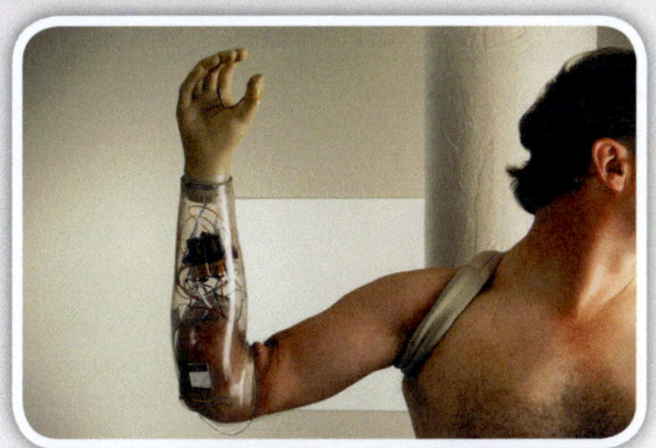

2 **Imagine a robot that explores the bottom of the ocean. What two features of a fish might the robot mimic?**

1. _________________________________
2. _________________________________

Lesson 2

What is the design process?

3 **Circle the goal of the design process.**

a) To redesign a prototype.
b) To communicate ideas to others.
c) To find a solution to a problem.

4 _________________ means to record what you've learned throughout the design process.

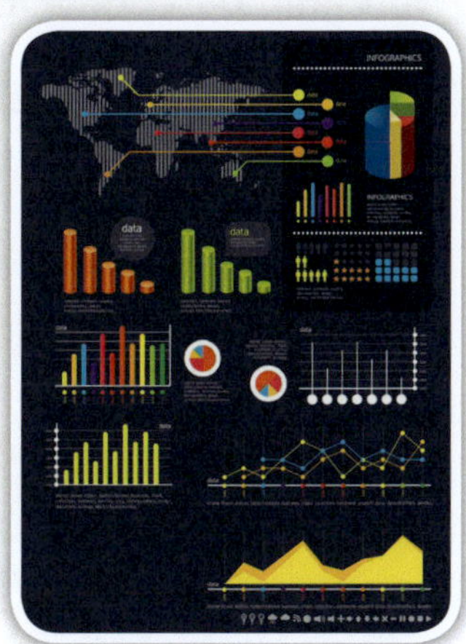

▷ **Got it? Quiz**   ▷ **Got it? Self Assessment**

Unit 2 Survival and Extinction

How do animals adapt to survive?

- that animals can live in different environments because of adaptations.
- how scientists use fossils to learn about the past.

1 Look and match the animal adaptations to their functions.

Humps enable some animals to store fat for periods of time when resources are scarce.

Large eyes on the front of their heads can help animals gauge distance for attacking prey with precision.

Thick fur provides insulation for animals living in cold climates.

2 With a partner, think of another animal that has an adaptation with a specific function. Describe it to the rest of the class.

3 What might a scientist conclude about a dinosaur that left fossil footprints of webbed feet? With a partner, discuss and circle the answer.

1. It could climb trees.

2. It could swim.

3. It could run fast.

Think!

What do you think are some advantages of having hard body coverings?

Lesson 1 · How do adaptations help animals?

1 Look and, with a partner, circle the parts of each animal that are adapted for swimming.

sea lion

duck

bluefin tuna

2 Read and underline four ways animals respond to a sudden threat.

Animal Adaptations

When there is a sudden threat, such as an attack by a predator, an animal such as the blue-ringed octopus may respond by changing color. Animals have both physical and behavioral **adaptations** that help them survive. The blue-ringed octopus is usually pale, but, when a predator approaches, it turns a bright yellow color with blue rings. Other animals might respond by running, flying, or using poison.

Changes in an environment, such as an increase in the salt content of the oceans over long periods of time, are too slow to affect individual animals. Animal species, like plant species, change over time to adapt to such slow changes in their environment.

3 With a partner, complete the table with a possible function of each adaptation.

Adaption	Function
poisonous bite	
large eyes	
long legs	

4 **Why do birds lay their eggs in the spring? Read and discuss with a partner.**

Life Cycle Variations

Different animals have different life cycles, which help them survive in their environment. For example, many birds lay their eggs in the spring when the weather warms up. The young hatchlings then have several months of good weather when food is available. This makes it easier for them to complete their growth.

At-Home Lab

Swimming Birds
With your fingers spread apart, move your hand through a tub full of water. With your fingers still spread apart, wrap plastic around your hand. Move it through the water again. Ducks have webbing on their feet. Consider how this adaptation helps a duck.

5 **Read and look. Then circle each adaptation in the pictures. In your notebook, write what purpose each one serves.**

Physical Characteristics

Physical adaptations help animals survive, too. Useful changes in the body parts of an animal are called structural adaptations. For example, animals that hunt tend to have eyes on the fronts of their heads. This makes them better at telling how far away their prey is, and they can pounce or swoop with precision. Animals that are hunted often have eyes on the sides of their heads because that helps them see where a predator might be coming from.

Individual animals do not develop structural adaptations. Instead, animal species develop physical adaptations through the process of natural selection, which also acts on plant species. Natural selection affects all species of living things.

gecko foot

sea urchin spines

okapi tongue

6 Look and, with a partner, say what is similar and what is different about the way the two birds use structural adaptations to find their food.

finch

owl

7 Why might a species not be able to survive? Read and circle all correct answers.

Extinction

A species cannot survive if it does not adapt to changes or move to a new environment. Some species cannot move to a new environment. For example, plants cannot pull themselves up by their roots and walk to another place. Also, the changes may be so widespread that there is nowhere left to move. If a species does not adapt to harmful changes or move away from them, its population will decrease. When a species has no members left that are alive, it becomes an **extinct species**.

a) If it is unable to adapt to environmental changes.

b) If it is unable to move or if there is nowhere for it to go.

c) If it is able to find a new habitat.

d) If it can adapt to the new environment.

8 With a partner, use the Internet to research information about what happened to the dodo species. Write your findings in note form.

 Read. Then look and circle _T_ (true) or _F_ (false).

Behavioral Adaptations

Were you born knowing how to build a house? That is impossible! Atlantic ghost crabs, though, are born knowing how to dig deep holes in the sandy beaches where they live. This behavior is due to genes passed from parent to offspring.

Behavioral adaptations are inherited behaviors that help animals survive. Behavioral adaptations are sometimes called **instincts**. They affect how an animal behaves around other animals. Some animals, like the ghost crab, have an instinct to burrow into the ground to hide from predators, such as shore birds.

Atlantic ghost crab

Not all behaviors are instincts. Some behaviors are learned by trial and error or as a result of training. For example, lion cubs learn to hunt by watching their parents and other animals. A lion cub learns to pounce on its prey by pouncing on its mother's twitching tail. When a zebra is separated from the herd, the adult lions will chase it toward a group of lions that are hiding. The lions will then pounce on their prey. The cub learns these behaviors over time.

1. The Atlantic ghost crab had to learn how to dig a hole. T / F

2. It learned to dig holes from watching its parents dig. T / F

3. It has natural digging instincts. T / F

4. It needs a tool to dig. T / F

10 **Why might lion cubs not need to know how to hunt when they're born? Discuss in small groups.**

11 **Read and complete the answers. Then match the answers with their questions.**

Seasonal Changes

In places with cold winters, there is little food for part of the year. Some animals deal with this food shortage by migrating, or moving. In spring and summer, Canada geese live in Canada and the northern United States. They **migrate** south to escape cold winter weather and to find food.

The snowshoe hare is brown in the summer and white in the winter. This makes it harder for predators to see.

Another type of seasonal behavior is hibernation. **Hibernation** is a state of inactivity that occurs in some animals when there is a shortage of food. These animals slow down or become inactive to conserve energy. Some mammals hibernate. Some reptiles and amphibians enter a state like hibernation.

1. How can animals deal with food shortages?

2. Which animals migrate for survival purposes?

3. What happens to animals when they hibernate?

When there is little ________________, some animals migrate or hibernate.

When animals ________________, they slow down or become inactive to conserve energy.

Canadian geese ________________ south when it gets cold. There, they can find food.

12 **What effect could less winter snow have on the fur of snowshoe hares over many generations? Discuss as a class.**

Think!

The central shaft in a bird's feather is hollow. What advantage might this give the bird?

At-Home Lab

Do research on an animal that migrates. Research where it goes and how long it stays in each place. Find out how far it travels. Present your findings to the class.

Lesson 2 · What can fossils tell us?

1 What clues does this fossil give about what sort of food this dinosaur ate? Mark (✓) and discuss with a partner.

☐ Its big, sharp teeth suggest it could eat meat.

☐ It looks as though it could eat only plants.

☐ The size of its mouth suggests it could eat large amounts of food.

☐ It looks like it could only eat leaves from bushes.

2 Read and underline what a paleontologist does. Then discuss with a partner what paleontologists can conclude about sauropods by studying their skulls and teeth.

Windows on the Past

Scientists study **fossils** to learn about plants, animals, and environments of the past. A scientist who studies fossils is called a **paleontologist**.

Some fossils show us what extinct organisms looked like and how they lived. For example, paleontologists have found skulls from dinosaurs called sauropods. **Sauropods** were a group of dinosaurs that had small heads, long necks, and enormous bodies. They ate plants and may have used their long necks to reach tall trees.

This illustration shows what one kind of sauropod may have looked like.

Because of their size, these dinosaurs needed huge amounts of food. However, their small heads meant that they could only take small bites. How did they get enough to eat? Paleontologists studied the dinosaurs' skulls and the shape of their teeth. They concluded that some sauropods may have swallowed their food without chewing it. Doing this allowed the dinosaurs to get food to their stomachs more quickly and take more bites.

3 **Read and circle *T* (true) or *F* (false). Then correct the false statements in your notebook.**

Fossils and Living Organisms

One way paleontologists learn about extinct plants and animals is by comparing them with plants and animals that exist today. For example, fossils show that some dinosaurs called **hadrosaurs** had large, hollow crests on their heads. Paleontologists have different ideas about the purposes of these crests. One idea came from comparing hadrosaurs to birds called peacocks. Male peacocks have large, brightly colored tails. They use their tails to attract mates. Paleontologists hypothesize that hadrosaurs may have used their crests in the same way.

Fossils can also show how plants and animals have changed over time. Many living things today are related to plants and animals from the past. Fossils show that some extinct plant species looked a lot like modern plants. For example, compare the pictures of the horsetail fossil and the modern horsetail plant. Some horsetail plants of the past grew to the size of trees. Modern horsetail plants are much smaller. This suggests that the plants changed slowly over time.

1. Hadrosaurs may have used their crests in a similar way to modern peacocks.   **T / F**

2. Fossils show that extinct plants didn't resemble modern plants.   **T / F**

3. The horsetail plant is much larger now than in the past.   **T / F**

4. Paleontologists believe plants change quickly over time.   **T / F**

4 **Look and label *horsetail fossil* and *modern horsetail plant*. Then, with a partner, discuss what is similar and what is different about the plants.**

________________________   ________________________

5 Read. Record below what scientists discovered in Kansas and South Dakota. With a partner, say what these findings mean about how their environments have changed.

Fossils and the Environment

Fossils also show that Earth's environment has changed. Scientists in Kansas found the remains of sea animals called ammonites. They are related to modern squids, but they died out 65 million years ago. In South Dakota, scientists discovered fossils of giant sea turtles. These turtles lived about 70 million years ago, but are now extinct. These discoveries show that areas of these present-day states were once covered with water!

Kansas	South Dakota

6 What might happen if we use up all the fossil fuels currently available? Read and discuss with a partner.

Fossil Fuels

Did you know that fuels such as **coal** and **oil** are a kind of fossil? Most of these **fossil fuels** come from the remains of organisms that lived millions of years ago. It took millions of years for the remains to become coal or oil. At power plants, fossil fuels are used to produce the electricity that powers homes and businesses.

Go Green

Fossil Fuel Use

With a partner, list three things you and your classmates can do to reduce electricity use and conserve fossil fuels. Share your list with the class.

7 In small groups, research alternative energy sources we can use instead of fossil fuels. Write your findings in your notebook. Then present your ideas to the class.

8 **Read. Then underline two ways scientists can determine the age of fossils.**

Fossil Age

Scientists determine the age of fossils in two ways. Many fossils are located within layers of rock. Older layers of rock are under newer layers. Scientists can conclude that fossils found in deeper layers are older than those in layers above. Scientists can also determine the age of fossils by examining how quickly certain materials in the fossils change. These materials change at steady rates after a plant or animal dies. Scientists can measure these materials in a fossil to determine how long ago the organism died.

Geologic Time Scale

Scientists have used information about the ages of fossils and rocks to make a timeline of the history of Earth. This timeline is called the geologic time scale. When they draw the scale, scientists place the earliest time span at the bottom. They put the most recent time span at the top. This matches the way rock layers of different ages are arranged. The time scale helps scientists show when different animals, including the dinosaurs, existed.

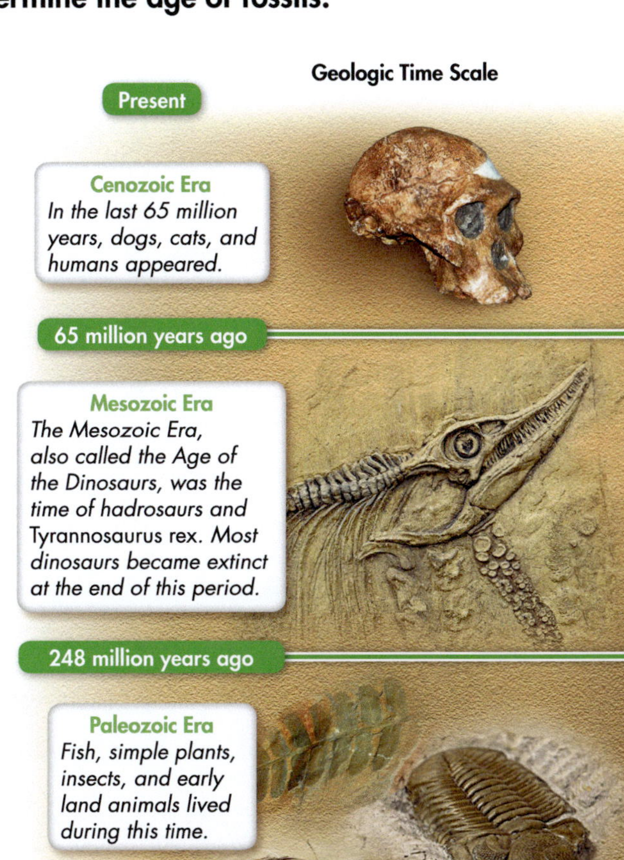

9 **Circle the correct era.**

1. Sea turtle fossils of South Dakota

   Cenozoic Era   Mesozoic Era   Paleozoic Era   Precambrian

2. Fossilized remains of bacteria

   Cenozoic Era   Mesozoic Era   Paleozoic Era   Precambrian

3. Skeleton of an early human

   Cenozoic Era   Mesozoic Era   Paleozoic Era   Precambrian

Let's Investigate! Lab

Which bird beak can crush seeds?

1. Make a model of a heron's beak by gluing 2 craft sticks to a clothespin. Use the other clothespin as a model of a cardinal's beak. Use pieces of straws as models of seeds.
2. Use the heron's beak to pick up a seed. Does the beak crush the seed? Try 5 times and record *yes* or *no*.
3. Repeat with the cardinal's beak. Record.
4. Which bird's beak crushes the seeds? _______________

Bird	Try 1	Try 2	Try 3	Try 4	Try 5
Heron					
Cardinal					

How do animals adapt to survive?

Lesson 1

How do adaptations help animals?

1 Label each animal adaptation *P* (physical) or *B* (behavioral).

☐ hunting prey ☐ poison ☐ thick fur ☐ burrowing into the ground

2 Look and label. Then circle the extinct species.

sea lion　　snowshoe hare　　dodo

Lesson 2

What can fossils tell us?

3 Mark (✓) all the true statements about fossils.

☐ They help scientists discover how environments have changed.

☐ They show us what extinct organisms looked like.

☐ They tell how species learned to hunt in the past.

☐ They show how plants and animals have changed over time.

4 Circle the correct answer.
What does the fossil of the horsetail plant reveal to scientists?
a) That the plant changed slowly over time.
b) That the plant is very different from the modern day horsetail.
c) That the plant is identical to the modern day horsetail.

▷ **Got it? Quiz**　　▷ **Got it? Self Assessment**　　Unit 2　**27**

Unit 3 — Body Systems and Function

How does my body work?

1 Look and label each body system.

circulatory system   respiratory system   nervous system

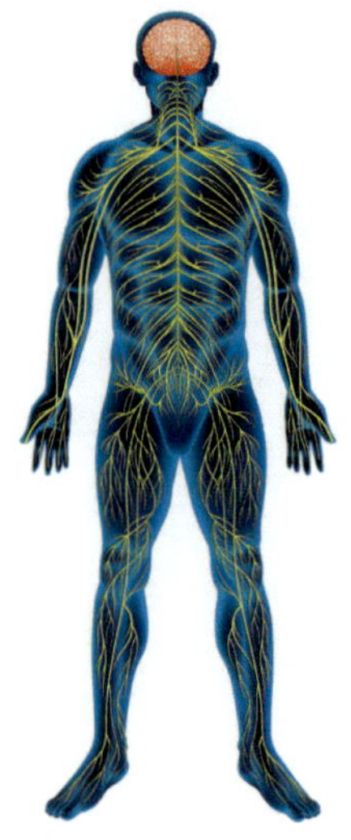

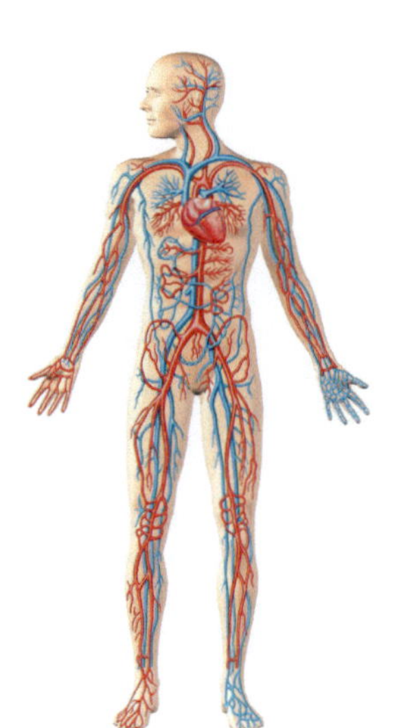

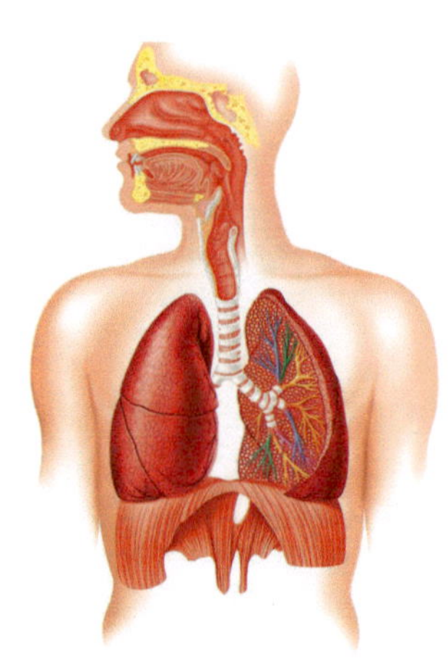

_________________   _________________   _________________

2 With a partner, decide which body system is involved in each situation. Write *CS* (circulatory system), *RS* (respiratory system), or *NS* (nervous system).

1. You inflate a balloon. _______
2. You burn your fingers and it hurts. _______
3. You feel your heartbeat on your neck or wrist. _______

3 With a partner, discuss what you think the role of each body system is.

Think!

How does your skin protect your body?

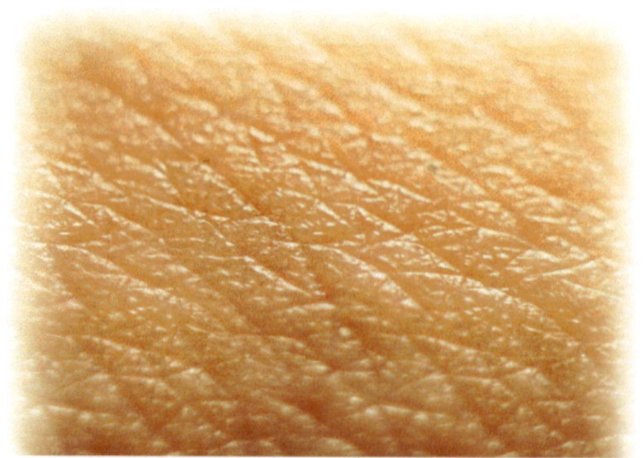

Lesson 1 · What is the circulatory system?

1 **Read, look, and, with a partner, describe where your heart is in your body. Then complete the statements.**

Cells to Organs

The smallest part of your body that is alive is a cell. Cells are the basic units of all living things. The tiniest organisms are made of only single cells. Larger organisms are made of many cells, maybe trillions of them.

Have you ever noticed that teamwork is a great way to get work done? Cells in larger organisms often work together in tissues. A **tissue** is a group of the same kind of cells that work together to do the same job.

Tissues join with other types of tissues to form organs. An **organ** is a group of different tissues that join together into one structure. These tissues work together to do one main job in the body. Your heart, eyes, ears, and stomach are all organs. Organs that work together are called an organ system.
A **system** is a set of things that work together as a whole.

Key Words

- tissue
- organ
- system
- circulatory system
- artery
- capillary
- vein
- heart

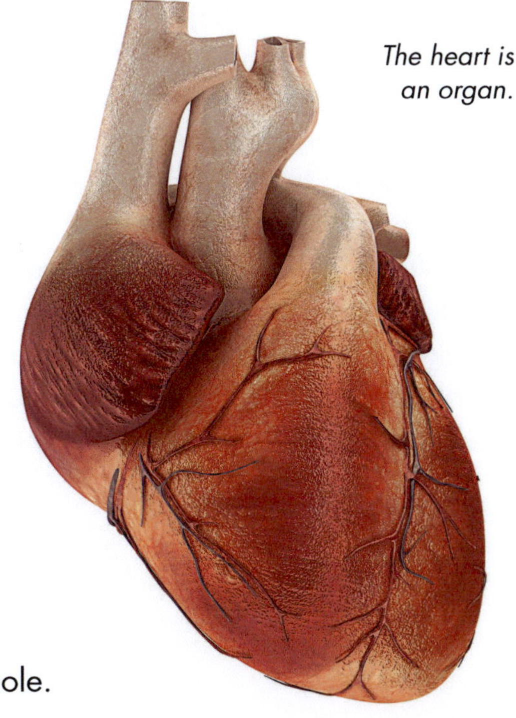

The heart is an organ.

1. Cells work together to form ________________________.

2. Tissues join with other tissues to form ________________________.

3. Organs work together to form an ________________________.

2 **Read and underline what the circulatory system includes. Then circle three kinds of blood vessels.**

Circulatory System

The **circulatory system** moves blood through the body. It includes the heart, blood, and blood vessels. Blood vessels are like highways for blood cells. The three kinds of blood vessels are arteries, capillaries, and veins. Most arteries carry blood with lots of oxygen.

3 **Read, look, and circle the organ that pumps blood through your body. Then, with the class, discuss the function of valves.**

Arteries, Capillaries, and Veins

Arteries are blood vessels that carry blood away from the heart to other parts of the body. Arteries have thick, muscular walls. These walls stretch as the heart pushes blood through them. Arteries branch many times into smaller and smaller tubes. The smallest arteries branch to become capillaries.

A **capillary** is the smallest kind of blood vessel. Capillary walls are only one cell thick. Oxygen and nutrients move from the blood in your capillaries through the thin walls to your body's cells. Carbon dioxide and other wastes move from cells to the blood in the capillaries.

Capillaries join together to form your smallest veins. **Veins** are blood vessels that transport blood toward the heart. Tiny veins join many times to form larger veins.

Unlike arteries and capillaries, veins have valves. Valves are flaps that act like doors to keep blood moving in only one direction. Valves open to let blood flow to the heart. They close if the blood starts flowing away from the heart.

4 **Read and fill in the blanks.**

veins	capillaries	arteries	toward	away from

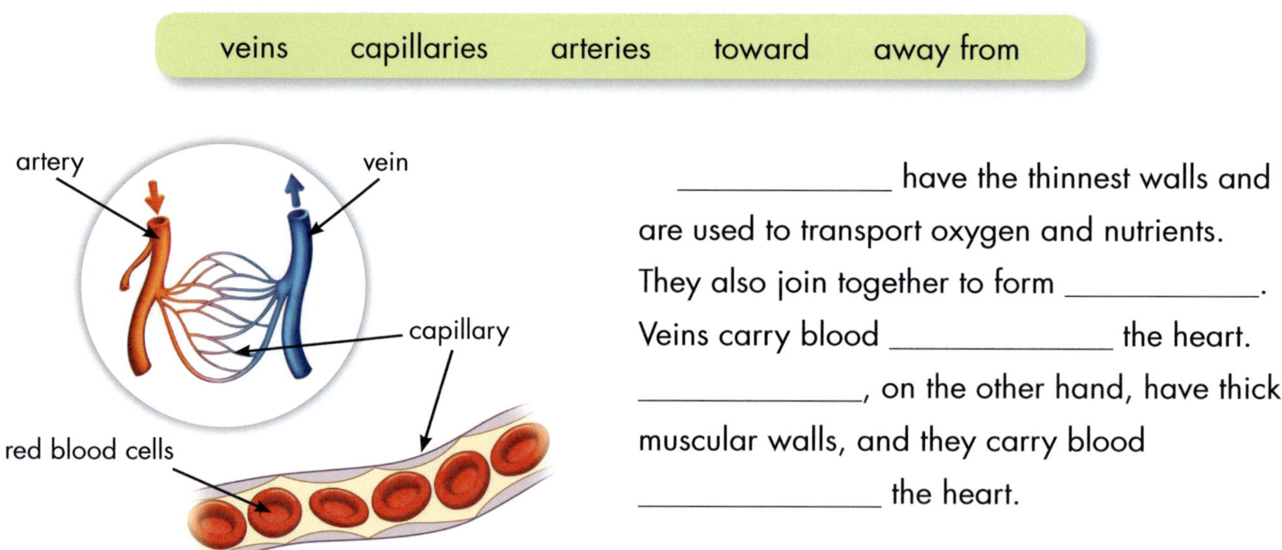

_______________ have the thinnest walls and are used to transport oxygen and nutrients. They also join together to form _______________. Veins carry blood _______________ the heart. _______________, on the other hand, have thick muscular walls, and they carry blood _______________ the heart.

Veins have thinner walls than arteries but thicker walls than capillaries.
Some capillaries are so narrow that red blood cells must flow through them single file.

5 **Read, look, and number the chambers of the heart (1–4) to show the order in which blood flows.**

Blood Flow through the Heart

The **heart** is a muscular organ that pumps blood throughout your body. Your heart is divided into right and left sides. Each side of the heart has two chambers and works as a separate pump to send blood along a different path. Blood enters the heart in the right atrium. Next, it flows to the right ventricle. The right ventricle pumps blood to the lungs. In the lungs, the blood gets oxygen and gets rid of carbon dioxide. Then blood returns from the lungs and flows into the left atrium. Finally, blood flows to the left ventricle. The left ventricle pumps the oxygen-rich blood through arteries to the entire body.

Parts of the Heart

Right Atrium
The right atrium relaxes and fills with blood carrying wastes and carbon dioxide from body cells. Then it contracts, squeezing blood into the right ventricle.

Left Atrium
Blood flows from the lungs into the left atrium. The left atrium squeezes blood into the left ventricle.

Right Ventricle
The right ventricle contracts, pumping blood into an artery leading to the lungs, where the blood can exchange carbon dioxide for oxygen.

Left Ventricle
The left ventricle pumps oxygen-rich blood away from the heart into your body's largest artery, called the aorta. From there, smaller arteries branch off as blood rushes to the body's cells.

In this drawing, the veins from the lungs are colored red because they contain oxygen-rich blood. Arteries going to the lungs are colored blue because they contain blood with less oxygen.

Think!

Why is it important that the heart pumps oxygen-rich blood through the body?

Lesson 2 · What is the respiratory system?

- respiratory system
- inhale
- diaphragm
- lungs
- exhale
- trachea
- bronchioles
- air sacs

1 Look and, with a partner, discuss how this man is using his respiratory system.

2 Read and underline two ways that your chest makes room for the air you breathe. Then complete the statement for each diagram.

The Respiratory System

Take a long, slow breath. Can you feel your respiratory system at work? The **respiratory system** is the system that helps you breathe. You take in air through your nose and mouth. Several muscles work together when you breathe. When you **inhale**, or breathe in, a dome-shaped muscle called the **diaphragm** moves down, making more space in your chest for air. Your rib muscles may also pull your rib cage up and out, making still more space. Air quickly rushes into your lungs and fills the space. The **lungs** are organs that help the body exchange oxygen and carbon dioxide with the air outside the body. When you **exhale**, or breathe out, your diaphragm and rib muscles relax, move up, and push air out of the lungs.

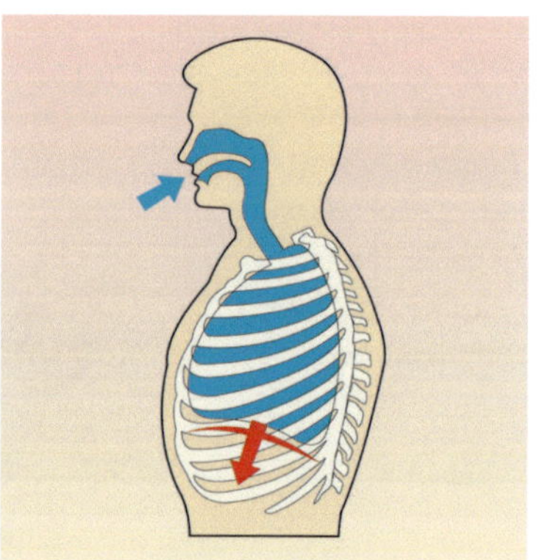

The diaphragm moves down when you ____________________.

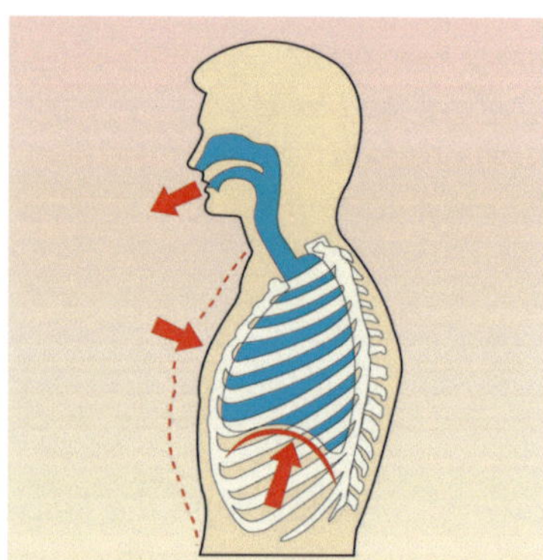

The diaphragm moves up when you ____________________.

Let's Explore! Lab

3 **Read and number the parts of the respiratory system to sequence how air passes through it.**

Parts of the Respiratory System

When you breathe, air comes in through your nose or mouth. Hairs and a layer of mucus in the nose trap dust, germs, and other things that may be in the air. Mucus is a sticky, thick fluid. Many parts of the respiratory system are coated with mucus.

From the nose, air passes through the nasal cavities. The nasal cavities warm and moisten the air. Then the air moves to the back of the throat and into the larynx. The larynx contains the vocal cords, where the voice is produced. The sound of your voice is the result of your breath making the vocal cords vibrate. When muscles stretch the vocal cords tighter, your voice gets a higher pitch.

From the larynx, a tube called the **trachea** carries air to the lungs. The trachea leads to two branches called bronchi that go into the lungs. In the lungs, the bronchi branch into smaller and smaller tubes called **bronchioles**.

The bronchioles end in clusters of tiny, thin-walled air sacs in the lungs. The **air sacs** are where oxygen enters the blood and carbon dioxide leaves the blood.

- ☐ air sacs
- ☐ bronchioles
- ☐ nasal cavities
- ☐ nose
- ☐ larynx
- ☐ trachea
- ☐ back of throat
- ☐ bronchi

Flash Lab

Slouch forward in your chair. Take a deep breath and then exhale. Now sit up straight. Take a deep breath and then exhale. In which position could you take a deeper breath? Explain.

Think!

Your left lung is smaller than your right lung. Why do you think this is so?

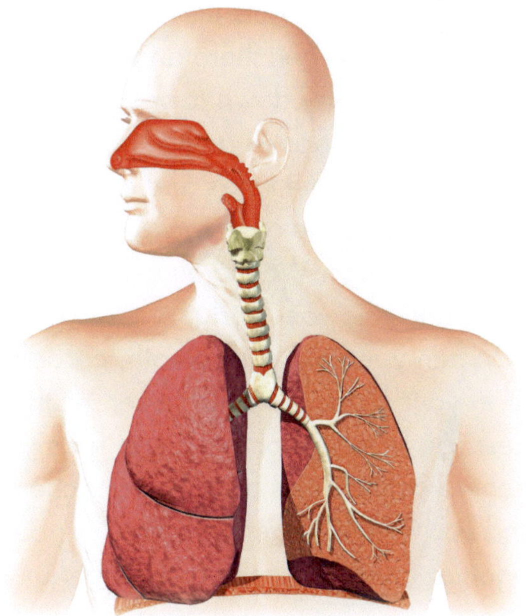

4 **Look and draw the path the air takes from your nose to your lungs.**

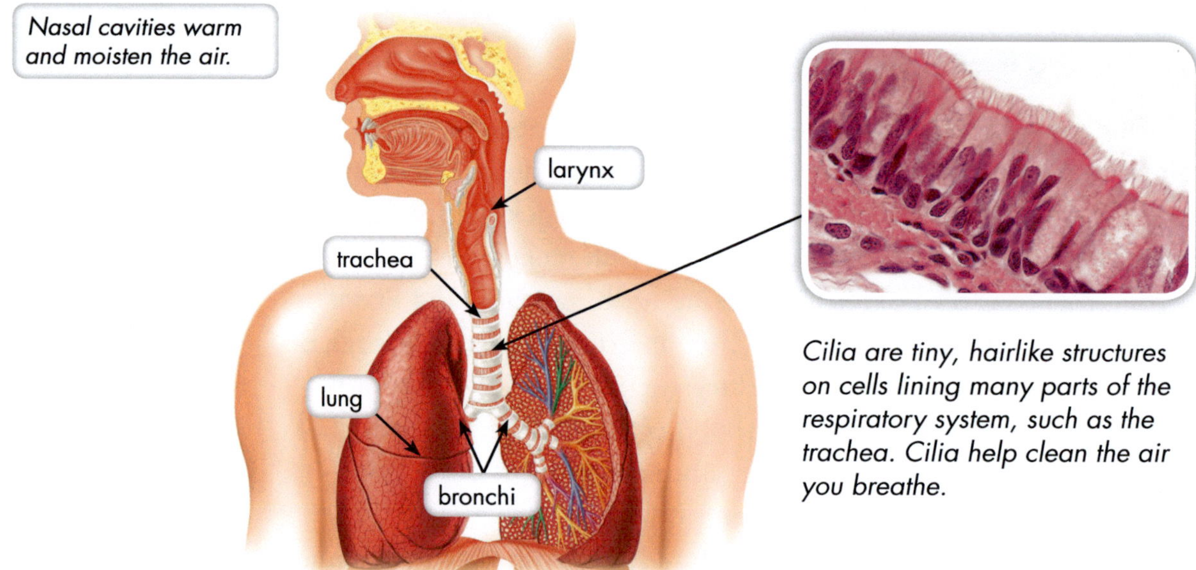

Cilia are tiny, hairlike structures on cells lining many parts of the respiratory system, such as the trachea. Cilia help clean the air you breathe.

5 **Read and, with a partner, summarize how oxygen is taken to your cells. Then, draw arrows on the red and blue blood vessels to show the direction of blood flow.**

Getting Oxygen to Cells

All of your cells need oxygen. You have a respiratory system and a circulatory system that work together to get oxygen to your cells. Oxygen enters your body when you inhale. Your respiratory system gets the oxygen as far as the tiny air sacs inside your chest. The blood picks up the oxygen there and carries it to your heart, where it is pumped to all of your cells—all the way down to your toes!

Two things happen at the same time in the air sacs. Oxygen leaves the lungs and enters the blood. Carbon dioxide moves the other way. It leaves the blood and enters the lungs. When you exhale, the extra carbon dioxide leaves your body.

When you hold your breath, carbon dioxide builds up in your blood. Your brain senses this. It sends a message to the diaphragm and rib muscles telling them to contract. As a result, your chest expands and you inhale. Several systems of your body work together to make sure your cells get oxygen.

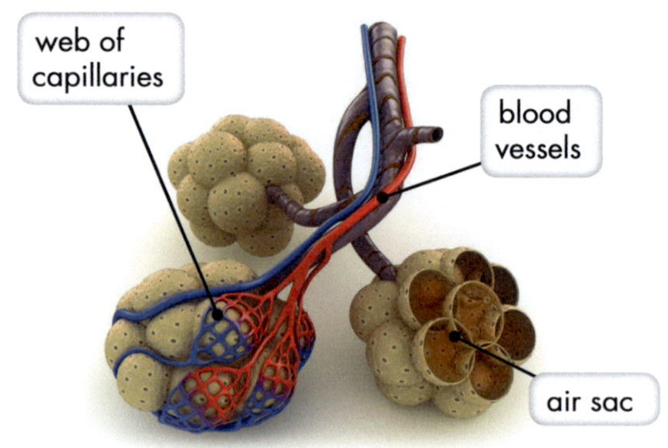

Air sacs increase the surface area of the lungs. This means more blood vessels can exchange oxygen and carbon dioxide.

 | **Lesson 2** Check | **Got it?** 60-Second Video

Lesson 3 · What is the nervous system?

1 Read, look, and, with a partner, say how the juggler is using his body systems.

Nervous System

In order to stay healthy, comfortable, and safe, your body needs information about your environment. It needs to know whether it is too cold or too hot, whether you are sitting down or standing up, and whether something hurts. Your body also needs to react appropriately, based on this information.

The system that receives information from your environment and controls how you react to it is the **nervous system**. The nervous system tells you what is going on in the world around you. It also tells your muscles how to contract to move your bones. The nervous system includes sense organs, nerves, the spinal cord, and the **brain**.

2 Read, look, and number each body part.

Sense and Sense Organs

You use your senses to know what is happening around you. The nervous system is constantly collecting information both inside and outside your body. It allows you to speak, think, taste, hear, and see. It helps the body stay balanced by processing and responding to the information it receives.

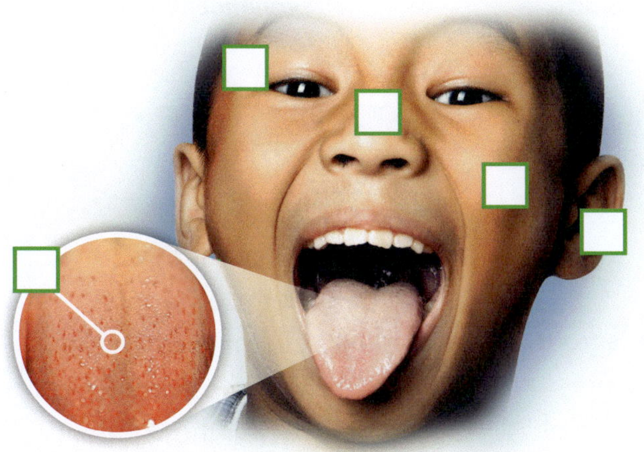

1. These have parts that sense light and send signals to the brain.

2. These have sensors that detect vibrations in sound waves. They also have sensors that help you control your balance.

3. The sense organ here responds to chemicals in odors. Signals from this organ are read by the brain.

5. Taste buds are small sense receptors located here.

4. Special sensors in the skin help you feel texture, changes in temperature, and sometimes pain.

3 Read and underline the parts that make up a neuron. Then draw an arrow on the neuron showing the direction a message travels through an axon.

Nerves

Nerve cells are also called neurons. Neurons pass messages throughout your body. Neurons are made of three parts: a cell body, an axon, and one or more dendrites. The cell body is the main part of the neuron. Dendrites receive messages from other neurons. The axon sends messages from the cell body to other neurons.

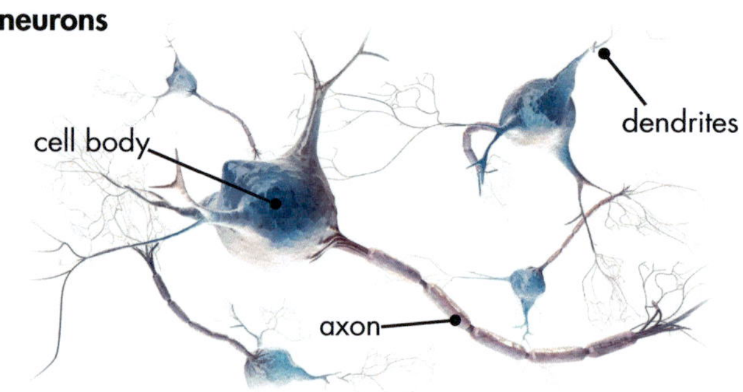

Messages can only travel in one direction between neurons. Most messages travel along neurons to the brain, which controls almost everything you experience and do. The brain interprets the message and responds by sending messages through neurons to different parts of the body, telling them to act.

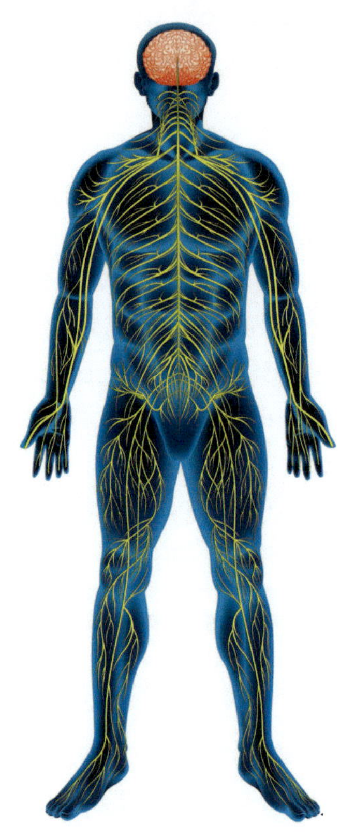

The nervous system is a system of structures that send information throughout the body.

4 Read and underline the function of your spinal cord. With the class, discuss activities you could not do if you injured your spinal cord.

Spinal Cord

Another important part of your nervous system is your spinal cord. Many messages received and sent by the brain pass through your spinal cord. This long bundle of nerves runs down your back and is protected by your backbone. Parts of the spinal cord carry messages to the brain. Others carry messages from the brain.

5 **Read and, with a partner, say the corresponding part of the brain that is used for each action.**

Brain Functions

Performing tasks, such as remembering, pretending, and feeling, are functions of the brain. So are running, playing games, and listening to music. The **brain** is the main organ, or control center, of the nervous system.

Your brain is made up of three major parts. The largest part of your brain is the cerebrum. This part of your brain learns, reasons, decides, stores memories, and feels fear and joy. Another part of your brain is the cerebellum. The cerebellum controls balance and posture. A third part of your brain is the brain stem. Your brain stem controls your blood pressure, heartbeat, breathing, and digestion.

1. Feeling happy.
2. Walking on a tightrope.
3. Taking a deep breath.
4. Remembering your childhood.

6 **Read and, in your notebook, list two more voluntary actions and involuntary actions. Compare with a partner.**

Voluntary Actions

A major function of your nervous system is to control voluntary actions. Voluntary actions are actions you decide to do, such as chewing, walking, or talking. The part of your brain that controls voluntary actions is the cerebrum.

Involuntary Actions

Another major function of your nervous system is to control involuntary actions. You do not need to think about these. Your brain stem controls some involuntary actions, such as the beating of your heart.

Some messages that the body receives do not pass to the brain at all. One example is the response of your body when you touch your hand to a hot surface. The response to that action is a reflex, a response that happens automatically without the brain "thinking" about it.

Let's Investigate!

How much air can you exhale?

1. Lay a trash bag over the top of a desk or table. Remove as many wrinkles as possible and tape down the edges.
2. Pour about 50 mL of bubble solution onto the bag.
3. Spread the solution around on the bag with a ruler. Dip a straw in the jar of bubble solution. Rest the straw on the wet bag. Take a deep breath and slowly breathe as much breath as you can into the straw. Observe a bubble forming.
4. Pop the bubble. Measure and record the diameter of the ring left on the bag. Use the Data Table to record the information for each student in your group.

Data Table		
Name of Student	Diameter of Ring (cm)	Volume of Air (L)

Volume Chart	
Diameter of Ring (cm)	Volume of Air (L)
14	0.7
15	0.9
16	1.1
17	1.3
18	1.5
19	1.8
20	2.1
21	2.4
22	2.8
23	3.2

Unit 3

Review

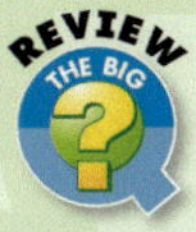

How does my body work?

Lesson 1

What is the circulatory system?

1 Read and circle the answer.

Blood entering the heart is pumped to the lungs before being pumped back out to the body. Why is this important?

a) Because blood picks up oxygen and releases carbon dioxide in the lungs.

b) Because blood picks up carbon dioxide and releases oxygen in the lungs.

c) Because blood picks up oxygen and carbon dioxide and releases them in the lungs.

Lesson 2

What is the respiratory system?

2 Circle the parts of the respiratory system.

ventricles trachea arteries bronchioles

nasal cavities veins nose lungs

Lesson 3

What is the nervous system?

3 Look and label each action *voluntary* or *involuntary*.

yawning ________________

eating ________________

Got it? Quiz   Got it? Self Assessment

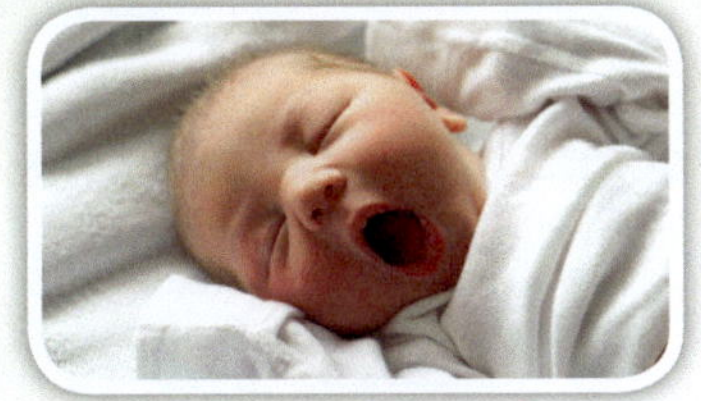

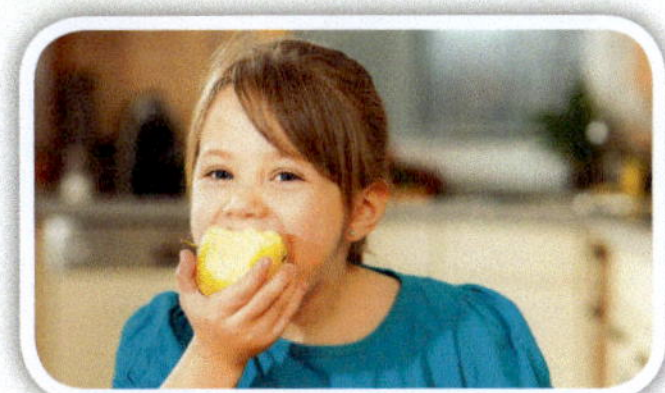

Water and Weather

How does water move through the environment?

1 Look and label.

moisture   air pressure   wind speed   thermometer   snow

__________   __________   __________   __________   __________

2 Look and mark (✓) what the pictures have in common.

sleet

snowflake

hail

[ ] a) They all show types of precipitation.

[ ] b) They all occur at high temperatures.

[ ] c) They all show types of frozen or semi-frozen water.

[ ] d) They all occur in dry conditions.

3 Look outside and, with a partner, describe the factors that determine the weather today.

Think!

Where do you think this water came from?

Lesson 1 · What is weather?

1 **Look at the map. Circle the area you might find clear skies. Draw an _X_ where you might find showers.**

2 **Read. Circle _T_ (true) or _F_ (false). Then, with a partner, correct the false statements.**

Weather

You probably know that a thermometer measures air temperature. But it takes more than a temperature reading to describe the weather. **Weather** is the state of the atmosphere, including its temperature, wind speed and direction, air pressure, moisture, amount of rain or snow, and other factors. Scientists called **meteorologists** study and predict weather. Meteorologists collect data using many tools to describe the current weather and to predict future weather. Weather observation stations collect almost all of the data automatically. Predicting the weather is very important for planning all sorts of activities, including farming, fishing, and outdoor concerts.

1. Weather is the temperature reading outside.   **T / F**

2. Meteorologists are scientists who predict the weather.   **T / F**

3. Meteorologists collect data manually in outside areas using tools.   **T / F**

4. It is important for agricultural workers to know weather predictions.   **T / F**

3 **With a partner, label the measuring tool, name two other tools for measuring the weather, and say what they measure.**

____

Let's Explore! Lab   Unit 4   **41**"

4 Read and circle three gases that make up the atmosphere. Then underline what happens to air pressure as you go higher into the atmosphere.

Barometric Pressure

When you look up on a clear day, you see a blue sky. You are really looking through 9,600 km (about 6,000 mi) of air. The blanket of air that surrounds Earth is its **atmosphere**. Like other matter, air has mass and takes up space.

Air is made up of a mixture of invisible gases. Over 3/4 of Earth's atmosphere is nitrogen. Most of the rest is oxygen, but small amounts of carbon dioxide gas are also present. The part of the atmosphere closest to Earth's surface contains water vapor. The amount of water vapor depends on time and place. For example, air over the ocean or a forest has more water vapor than air over a desert.

Gravity pulls the mass of the air in the atmosphere toward Earth's surface. The pushing force of the atmosphere is called **barometric pressure**. Air pushes with equal force in all directions. Many kilograms of gas are pressing down on your school building. They do not crush it because the air inside the building exerts pressure, too. Air pushing down is balanced by air pushing up and sideways. Air pressure decreases as you go higher in the atmosphere.

5 Look and circle the part of the mountain where the air pressure is the lowest. Discuss why with a partner.

Think!

If you take two readings from a barometer, one reading from the top of a tall building and the other at ground level, which reading is likely to be higher? Why?

6 What would happen if the air outside the hot air balloon were as hot as the air inside? Read, look, and discuss with the class.

Temperature

Air temperature also affects weather. As the sun warms Earth's surface, air that is in contact with the surface becomes warmer. As the air particles move farther apart, the air pushes down with less pressure. The warm air rises, causing an area of low pressure to form, and air from areas with higher pressure rushes in. If the air near Earth's surface cools, the particles in the air become more closely packed. This denser, cooler air pushes down with more pressure. An area of high pressure forms. Air from this area flows into lower-pressure areas. The temperature of the air also affects the type of precipitation—rain, snow, or sleet.

7 Read and underline what a jet stream is. What happens in Canada in the summer? Circle the correct answers.

Winds

Wind is air movement caused by differences in pressure. In general, air moves from areas of high pressure to areas of low pressure. When you let air out of a balloon, air rushes from inside the balloon where pressure is higher to where pressure is lower outside the balloon.

Wind speed and direction affect weather. Jet streams can affect local weather. A **jet stream** is a narrow band of high-speed wind. A polar jet stream blows from west to east high in the atmosphere over North America. In the winter, it can take cold air from the north as far south as Kentucky. In the summer, it takes warmer air north into Canada.

A wind is named based on the direction from which it blows. A north wind comes from the north and moves toward the south.

Meteorologists measure wind speed using an instrument called an anemometer.

a) A jet stream blows from west to east.
b) Cold air blows in from the south.
c) A jet stream blows warm air from the south.
d) Warm air blows from Canada to Kentucky.

8 Look and draw an arrow on the wind vane to represent a southeasterly wind. The direction of the arrow shows the direction the wind is coming from.

9 Read and mark (✓) all the true statements.

Water in the Atmosphere

Three other factors for determining weather are humidity, clouds, and precipitation. **Humidity** is the amount of water vapor in the air. The particles of water vapor are too small to be visible, but, when conditions are right, they can come together to form small water droplets and ice crystals. These droplets and crystals are bigger than water vapor particles and can reflect light from the sun. At this point, we can see the water as a cloud. If the droplets or crystals get large enough, they can fall to the ground as precipitation, such as rain or snow.

- ☐ If particles of water vapor come together to form droplets and crystals, we can see them as clouds.

- ☐ Snow falls when water droplets or crystals are small.

- ☐ Humidity, clouds, and precipitation are all related to water.

- ☐ We can see particles of water vapor in the air because they look like rain.

10 Look and, with a partner, describe what is happening in the picture. Use the words *humidity*, *clouds*, and *precipitation* in your explanation.

11 **Read, look, and draw arrows to show the motion of air in the room in winter.**

Circulation

Have you ever used an electric fan to cool a room in the summer? You can also use a fan to make a heater more efficient in the winter. The fan moves air around the room.

The wind may blow from different directions, but winds do follow some large-scale patterns over continents and the ocean. These patterns are determined by differences in temperature and pressure in different parts of the atmosphere. The large-scale movement of air is called circulation. ==Circulation== is the movement of air that redistributes heat on Earth.

For example, the ==trade winds== are a persistent pattern of winds that blow near the ==equator==. The warmest parts of our planet are near the equator. The air above this region becomes warm and rises, creating a low pressure zone. High in the atmosphere, this warm air travels away from the equator, cools down, and sinks.

In summer, a ceiling fan should blow the air downward. This makes sweat evaporate faster and people feel cooler.

In winter, a ceiling fan should draw the air upward. In a heated room, the hot air rises to the ceiling. The cold air blowing up pushes the hot air toward the walls and then down, where people can be warmed by it.

12 **Look at the route for a ship traveling from Europe to America and back. Draw arrows to show the direction the ship could take to save the most fuel. With a partner, explain the direction you chose.**

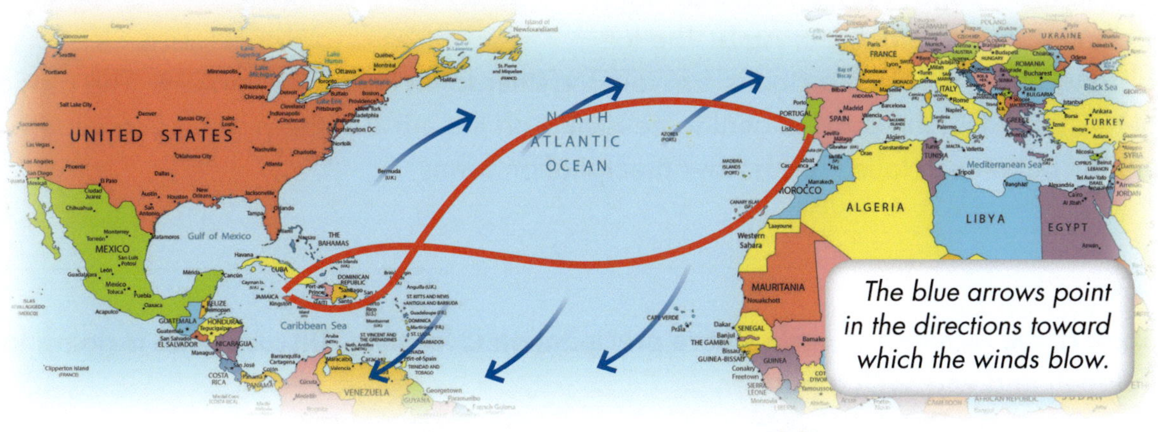

Lesson 2 · How do clouds and precipitation form?

1 Look and, with a partner, say what figures you can see in the clouds. As a class, discuss why clouds have different appearances.

2 Read and write the two substances that clouds are made of.

Water in the Air

Have you ever watched a cloud get larger? Have you tried to see shapes in the clouds? Clouds come in many shapes and sizes. Remember that clouds form when water vapor changes into tiny water droplets or ice crystals. Whether a cloud is made of water droplets or ice crystals depends partly on air temperature. The temperature of air high in the clouds is often much lower than the temperature of the air close to the ground. Even on summer days, many clouds are made of ice crystals. The ice crystals and water droplets in clouds can join together to make larger particles. The particles can get so large that the gravitational force acting on the mass of the particles can cause them to fall out of the cloud. This is how precipitation forms.

1. _______________________________   2. _______________________________

3 Look at the picture and circle where precipitation is forming. Discuss what is going to happen and why with a partner.

   ▷ **Let's Explore!** Lab

4 **Read and fill in the graphic organizer below.**

Precipitation

You may be surprised to learn that most rain in the United States starts as snow. The temperature of the air high above the ground is often below 0 °C. Clouds of ice crystals form in the cold air. The ice crystals grow larger until they start to fall as **snowflakes**. As they fall, the crystals sometimes stick to other crystals and become larger snowflakes. If the temperature of all the air between the cloud and the ground is less than 0 °C, the ice crystals will fall to the ground as snowflakes. The ice crystals from a cloud may change as they fall through different layers of air. If the ice crystals fall into air that is warmer than 0 °C, they will melt and fall as rain. If the air near the ground is very cold, the rain sometimes freezes before it hits the ground. The frozen raindrops are **sleet**.

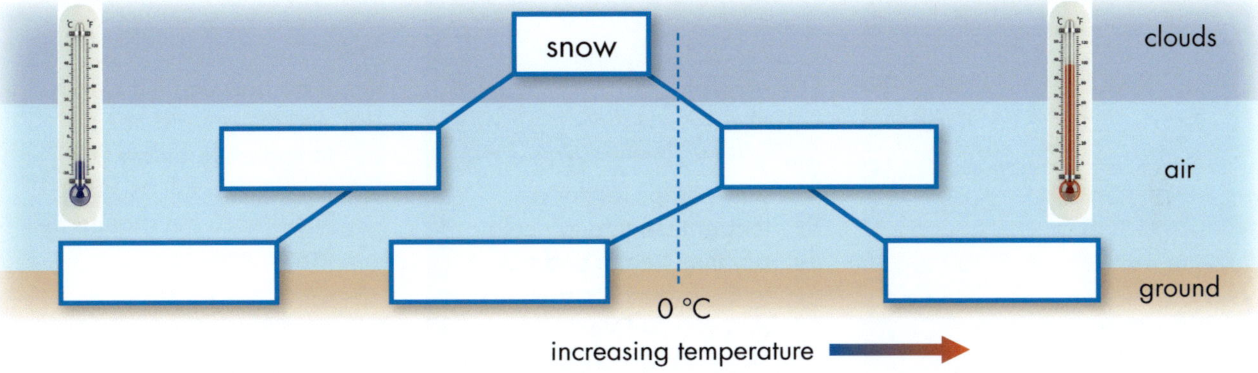

5 **What causes layers to form in hail? Read and discuss with a partner.**

Hail

If strong winds blow up through a thunderstorm cloud, raindrops are blown back up into the freezing air at the top of the cloud, creating a small piece of ice. As the ice is blown through the cloud many times, many layers of water freeze on it. Finally, it gets too heavy for the winds to carry it back up. This frozen precipitation that forms in layers is called **hail**. The hailstones fall to the ground.

I Will Know...   Unit 4   **47**

6 Look at the chart. With a partner, circle one way rain and sleet are alike and underline one way snow and sleet are alike.

7 **Read. Then, with a partner, circle the picture of the clouds that you can see outside today.**

Types of Clouds

Different cloud types form depending on the type of weather present. Clouds that form at different **altitudes** in the atmosphere have different names. Here are five common types of clouds.

1. Cirrus

High-level clouds form more than 6 km above the ground. This region overlaps the region for mid-altitude clouds. Cirrus are high-altitude clouds that are often thin, wispy, and white.

2. Cumulonimbus

Clouds that grow vertically have rising air inside them. The bases of these clouds may be as low as 1 km above the ground. The rising air may push the tops of these clouds higher than 12 km. Vertical clouds can cause thunderstorms.

3. Altocumulus

The bases of mid-level clouds are between 2 km and 7 km above the ground. Altocumulus clouds are mid-level clouds that look like small, puffy balls. The bottoms of the clouds can look dark because sunlight may not reach them.

4. Stratus

Low-level clouds are often seen less than 2 km above the ground. Stratus clouds are low-level clouds that cover the whole sky. They look dark because little sunlight gets through the layer of clouds.

5. Fog

Fog is a cloud at ground level. As air near the ground cools, water vapor condenses into tiny droplets and forms a cloud at or near the ground. As more droplets form and get larger, the fog appears thicker.

local 5-day weather forecast showing temperature, possible precipitation, sunshine, etc., as shown for "Monday" above

Let's Investigate!

How accurate are weather forecasts?

1. Look at the current 5-day weather forecast and record the forecasted high temperatures.
2. Check the weather report each day for the next 5 days. Record the actual high for the previous day.
3. Compare the forecasted data with the actual data. What was the largest difference between the forecast and actual temperatures? _______ °C.

Weather Report Predictions			
Day	Forecast High (°C)	Actual High (°C)	Difference Between Forecast and Actual (°C)
1			
2			
3			
4			
5			

   ▷ **Let's Investigate!** Lab

REVIEW THE BIG
How does water move through the environment?

What is weather?

1 **Read and circle the correct sentence ending.**
Weather is…

a) the temperature of the air at a given time and place.

b) the temperature and amount of moisture in the atmosphere at a given time and place.

c) the state of the atmosphere, including its temperature, wind speed and direction, air pressure, moisture, amount of rain or snow, and other factors.

2 **Look and write three factors that determine the weather shown.**

1. ______________________________
2. ______________________________
3. ______________________________

How do clouds and precipitation form?

3 **Circle what clouds are made of.**
a) snowflakes or sleet   b) water droplets or ice crystals   c) water vapor or snow

4 **Circle the type of precipitation that forms when ice particles get blown up through a thunderstorm cloud by a strong wind.**

snowflake

hail

sleet

Earth's Surface

THE BIG Q — **What kind of processes change Earth's surface?**

1 Read and match each type of energy with its description.

Renewable Energy	Nonrenewable Energy	Inexhaustible Energy

This type of energy comes from natural resources on Earth that will never run out. Wind is an example of this energy.

This type of energy is replaceable, but often only over long periods of time. Wood and manure are examples of this energy.

This is an irreplaceable energy source or a resource that cannot be replaced as quickly as people need. Nuclear energy is an example of this.

2 Look and write *R* (renewable energy), *N* (nonrenewable energy), or *I* (inexhaustible energy). Check answers with a partner.

3 With a partner, make a list of the top five energy sources you think are used in your neighborhood. Then compare with the class.

Think!

What forces might have shaped this rock formation?

Lesson 1 · How does Earth's surface change?

1 How did this part of Earth's surface change? Look and discuss with a partner.

2 Read and circle *T* (true) or *F* (false). Then correct the false statements orally with a partner.

Earth's Plates

The solid, rocky outer-most layer of Earth is called the **lithosphere**. The lithosphere covers all of Earth, but it is not a solid sheet of material. It is broken into several large sections and many smaller ones. A section of the lithosphere is called a **plate**. Several plates are larger than continents. A plate may include continents, parts of the ocean floor, or both. Earth's plates move slowly. They might push into each other, pull apart, or grind past each other.

1. The lithosphere covers part of Earth and comprises a number of plates.   **T / F**

2. Plates can include land and part of the ocean floor.   **T / F**

3. Plates move slowly and in opposite directions.   **T / F**

4. A plate is no larger than a single country.   **T / F**

3 Look at the map that shows the Earth's plates. Circle the plate that you live on.

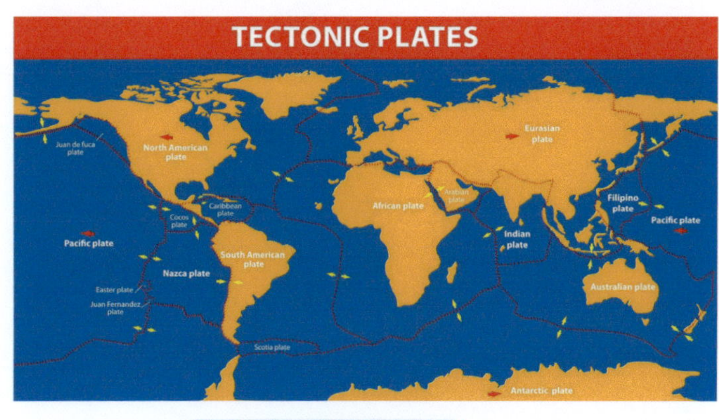

4 **Read and underline three types of plate boundary. Then look and label the type of plate boundary below. Check answers as a class.**

Changes over Time

Plate movements are small. Some plates move less than one centimeter per year. Some plates move as much as ten centimeters per year. Even so, these movements can cause big changes to Earth's surface. Some changes occur slowly over thousands or millions of years. These changes include mountains being built and valleys being formed. Some changes happen quickly in days or even minutes.

Generally, mountains form, earthquakes occur, and volcanoes erupt in certain places. These places are where plates meet. The edges where Earth's plates meet are called plate boundaries. A **converging boundary** occurs when two plates push into each other. A **spreading plate boundary** may form when plates move apart from each other. At a **sliding plate boundary**, two plates move past each other in opposite directions. You can see a crack in the land in some places where two plates meet. This crack is called a **fault**.

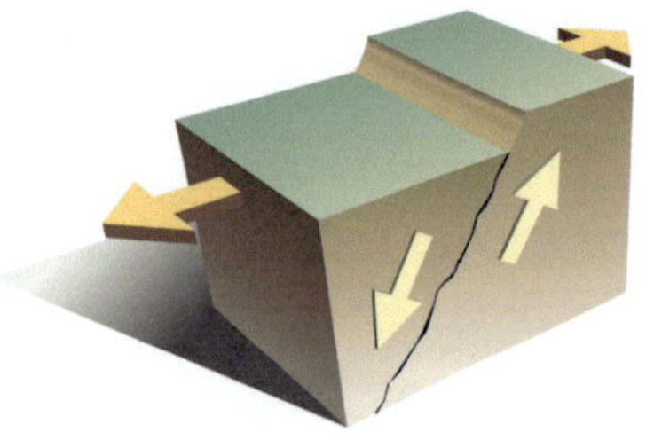

Faults may be seen where some plates meet.

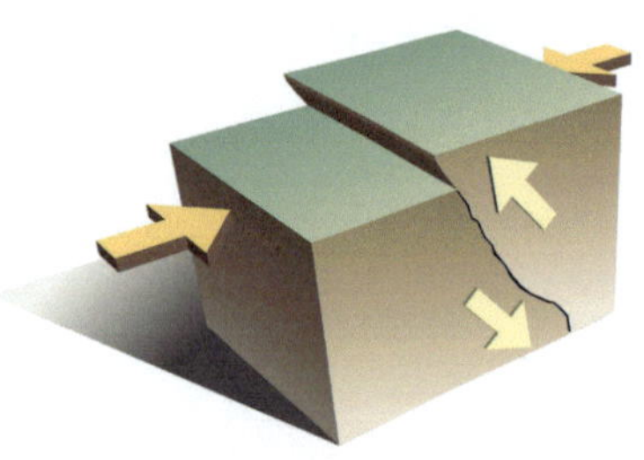

Mountains form where plates collide.

Volcanoes can form near plate boundaries.

I Will Know...

5 Read and, with a partner, describe what constructive and destructive forces are. Then research examples of changes brought about by both forces and make a list of each in your notebook.

Constructive and Destructive Forces

Many forces change Earth's surface. These changes mostly occur at plate boundaries. **Constructive forces** build new features on Earth's surface. Forces that wear away or tear down features are **destructive forces**.

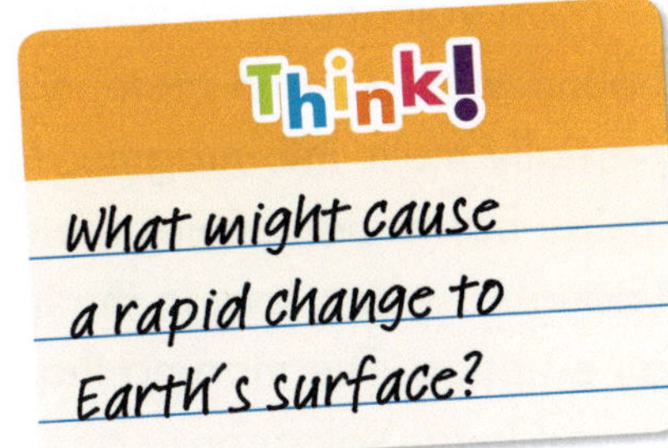

Mountains and Valleys

Constructive forces form new mountains and valleys. Mountains form when Earth's crust folds, tilts, and lifts as plates collide. New valleys can form at spreading plate boundaries, such as the one under the Atlantic Ocean. At this boundary, the ocean floor looks like a mountain range. It is called the Mid-Atlantic Ridge. The low area between the plates is a rift valley. It is slowly becoming wider.

6 Read and underline three examples of destructive forces.

Earthquakes

Earthquakes happen along faults. Faults are cracks in Earth's lithosphere where the surrounding rock has shifted. Faults can form anywhere. Earthquakes most often occur at faults along plate boundaries. The plates get "hung up" and lock in place. Eventually the plates jerk into a new position. This sudden movement and the resulting vibrations cause an earthquake. Plate movements often happen far below ground.

Energy released in an earthquake can destroy things quickly. It can cause landslides, which are downhill movements of large amounts of rock. Earthquakes under the ocean can cause tsunamis. Tsunamis are large waves of water that can crash into a coastline, destroy things, and wash away soil.

An earthquake can cause sections of road to crack, twist, or be completely destroyed.

7 **Read. Then fill in the blanks.**

Volcanoes

Most volcanoes on land form near converging plate boundaries. As one plate moves below another plate, rock partially melts into magma. Magma is a hot liquid material inside Earth. Sometimes the magma is forced to the surface through a weak spot in the lithosphere. This action is called an eruption. The magma that reaches Earth's surface is called lava. Volcanoes can do more than ooze fountains of lava. Gases, such as water vapor and carbon dioxide, are often mixed with the lava. Trapped gases can have enough pressure to blow apart the side of a volcano during an eruption. These trapped gases can push lava high into the air. While it is still in the air, this lava may cool into ash or rock.

Volcanoes can form on continents. They can also build from the ocean floor. A volcanic island forms when a volcano reaches the surface of the water. This is also a constructive process. The state of Hawaii is a string of islands formed in this way.

1. Volcanoes generally form close to _______________________ plate boundaries.

2. The hot liquid rock inside Earth is called _______________________.

3. An _______________________ occurs when magma is forced up through the lithosphere.

4. When magma reaches the Earth's surface, it is called _______________________, but, when this cools in the air, it can become _______________________ or _______________________.

8 **Is this volcano an example of a constructive or destructive force? Why? Discuss with the class.**

Flatten a ball of clay. Push the ends of it toward each other. Then pull the ends apart. Describe the result of each movement. How would Earth's surface have changed if the clay were Earth's plates?

   Lesson 1 Check   Got it?   60-Second Video

Lesson 2 · What are some energy resources?

1 Read and underline two examples of energy resources. Then look and, with a partner, say how you think turbines might use the energy in wind.

Energy Resources

A resource is something that will meet a need. An energy resource is something that will meet energy needs. The sun is an energy resource. It gives off energy in the forms of light and heat. People can use this light to make electricity. Trees are also an energy resource because they provide wood. Wood produces heat when it burns, so people can use it to heat their homes. Renewable, nonrenewable, and inexhaustible resources are three major types of natural resources.

2 Read and answer the questions below. Then, as a class, discuss what might be some disadvantages to inexhaustible resources.

Renewable and Inexhaustible Resources

Resources that can be replaced are **renewable resources**. Renewable energy resources include the wood from trees, leaves, food wastes, and even manure. These resources belong to a group of fuels called **biomass fuels**, which are made by living things or from recently living things. Some biomass, such as corn, can be turned into fuels that can run cars and trucks. Some garbage can even be used as biomass. By using garbage as a resource, less of it will be taken to landfills. One disadvantage of biomass is that burning it causes air pollution.

Sun, wind, moving water, and energy from inside Earth are inexhaustible resources. **Inexhaustible resources** will not run out.

1. Mark (✓) the statements that are true.

☐ Renewable resources can be obtained from living things.

☐ Some biomass can be turned into fuels for vehicles.

☐ Inexhaustible resources have the disadvantage of causing pollution.

☐ Inexhaustible resources can come from inside the Earth.

2. What are two effects of using biomass for energy?

a) _________________________________ b) _________________________________

3 **Read and label each energy type. Then number the pictures.**

Solar   Wind   Geothermal   Biomass

1. ________________________________

Trees provide wood. Wood is probably the first fuel that people ever used for both heat and light. Although wood is a renewable fuel, it takes time for new trees to grow. Also, burning wood increases the amount of carbon dioxide in the air, which contributes to global warming.

2. ________________________________

This is the motion of air. People have used this for energy for thousands of years. Windmill blades were connected to machines to grind grain and pump water. Today, turbines use this energy to spin generators that make electricity. This is an inexhaustible resource.

3. ________________________________

This energy comes from sunlight. A device called a solar cell changes this energy into electrical energy. When light hits the cell, an electric current is produced. Groups of solar cells form solar panels. Some homes, buildings, and cars have solar panels to provide energy.

4. ________________________________

This energy comes from the high temperature inside Earth. One way to get this energy is to pump water down deep holes into hot rock inside Earth. The hot rock heats the water or turns the water into steam, which rushes back to Earth's surface and can be used to make electricity.

Think!

Where does the energy in wood originally come from?

4 **Read and discuss with a partner how coal is formed. Then number the sequence for coal formation.**

Nonrenewable Resources

Energy resources that either cannot be replaced at all or cannot be replaced as fast as people use them are **nonrenewable resources**. Nuclear energy and fossil fuels are nonrenewable energy sources.

Nuclear Energy

Nuclear power plants often use a metal called uranium to heat water into steam. This steam is used to power generators that produce electricity. Nuclear power plants do not release pollution into the air. But their wastes are dangerous and must be stored in special places.

Fossil Fuels

Coal, oil, and natural gas are also nonrenewable energy sources. They are called fossil fuels because they are made from the remains of organisms that lived long ago. Coal forms from plants. Under certain conditions, layers of dead plants build up and form a material called peat. Peat gets buried and slowly changes into soft coal and then into hard coal. Today, coal fuels many electric power plants. Burning coal turns water into steam. Steam causes generators to spin and make electricity.

5 **Read. In small groups use the Internet to research the uses of oil and natural gas. Make a list for each in your notebook and present it to the class.**

Oil and Natural Gas

Unlike coal, oil forms from the remains of tiny sea organisms, not dead plants. Oil may be found beneath land or beneath the ocean floor. Drills make deep holes in Earth's surface to reach the oil.

Natural gas is often found where oil is found. The gas is usually pumped into pipelines. Pipelines carry the gas to storage tanks until it is needed.

Coal Formation

coal

plant life

peat

Lesson 3 · What is pollution?

1 Look and circle the items that may harm this environment. With a partner, explain how.

2 Read and underline two pollutants. Then number (1–3) to sequence each event.

Pollutants

When you ride in a car or brush your teeth, you are using natural resources. People must use natural resources to live. However, sometimes pollutants harm the planet. A **pollutant** is an unwanted substance added to the water, air, or soil. Pollutants result in pollution.

Sources of pollutants may be natural. A volcanic eruption can throw dust, ash, and other particles into the air. The particles can block sunlight for a long period of time. Without this sunlight, temperatures may drop and plants might die. People may also be sources of pollutants. Chemicals that people use to kill insects and weeds may harm other organisms.

sandstorm

☐ Dust and sand cover the ground.	☐ A sandstorm sends dust hundreds of miles away.	☐ Plants die.
☐ Plants might die without the sun's light.	☐ Dust and ash in the air block sunlight.	☐ A volcano erupts.

  ▷ **Let's Explore!** Lab

3 **Read and complete the activities below.**

Pollutants in Air

Some air pollution is caused by volcanoes and forest fires. But more and more often, air pollution results from the actions of people. Automobiles and coal-burning power plants release harmful chemicals into the air. These can be unhealthy to breathe, and they can also harm plants and the animals that depend on them.

Pollutants in Water

Water is sometimes polluted by factories that dump harmful chemicals into rivers, lakes, and oceans. These chemicals can harm or kill fish and other animals or plants. Oil is often transported in large tankers. If one leaks, oil gets into the water and can wash up onto shore. Oil spills can harm or kill animals and plants that live in the water and on the coast.

1. Write four sources of air pollutants.

   1. ________________ 2. ________________ 3. ________________ 4. ________________

2. Fill in the blanks.

| leak | ocean | chemicals | tankers | plants | animals | dumped |

Harmful ________________ that are ________________ into lakes, rivers, and the ________________ can harm life. Oil from ________________ that ________________ can also kill ________________ and fish or other ________________ that live near the water.

4 **Read and underline how dumps caused problems and circle how landfills can be used in a positive way. Then discuss with the class ways to use and reduce garbage.**

Pollutants in Soil

Soil becomes polluted by garbage, litter, and other solid waste. Garbage used to be put into open dumps. But rain that fell on them washed harmful chemicals into the soil. Today, most garbage is buried under the ground in **landfills**. Trees and grass can be planted on top of the ground, and people can use the land again. Sometimes people throw garbage onto the ground. This kind of pollution is called **litter**, which can be harmful to many animals and can spread germs.

Lesson 3 Check   Got it?   60-Second Video   Unit 5   **61**

Let's Investigate! Lab

How do forces affect Earth's surface?

1. Cut an index card into the size of a sponge. Tape it to the sponge. On the other side of the sponge, put a second sponge. Connect them with a rubber band.
2. Join 2 other sponges with a rubber band.
3. Put 2 desks about 10 cm apart. Put a pair of sponges on each side of the gap. Be sure to place the cardboard side down. Let the sponges hang over the edges of the desk.
4. Slowly, move the sponges together and record what you see.

____

____

____

____

____

REVIEW THE BIG ?

What kind of processes change Earth's surface?

How does Earth's surface change?

1 **Circle the answer.**
What kind of plate boundary might form a mountain?
a) a converging boundary
b) a spreading boundary
c) a sliding boundary

What are some energy resources?

2 Look and label each energy resource *renewable*, *nonrenewable*, or *inexhaustible*.

__________________   __________________   __________________

What is pollution?

3 Write an example of each.

An air pollutant   __________________________
A water pollutant   __________________________
A soil pollutant   __________________________

▷ **Got it? Quiz**   ▷ **Got it? Self Assessment**   Unit 5   **63**

Unit 6 — Earth and Space

How do objects move in space?

1 Look and complete the label for each picture.

ast ___ r ___ id   m ___ o ___   m ___ teo ___   c ___ me ___

2 What do all of these objects have in common? Discuss with a partner.

3 What is the sun? Circle and discuss with a partner.

a) It is a star and a hot ball of glowing gases.

b) It is a spherical, rocky mass.

c) It is a star at the center of our universe.

d) It is a mass of ice and dust in orbit around a larger star.

Think!

Why do you think the sun sometimes looks like a crescent?

Lesson 1 · What is a star?

- photosphere
- chromosphere
- corona
- sunspots
- prominence
- solar flare
- constellation

1 **Does the sun have a hard surface like Earth? Look and discuss with the class.**

2 **Read and circle *T* (true) or *F* (false). Correct the false statements orally with a partner.**

Stars

Stars are gigantic balls of very hot gases that give off radiation. The sun is a medium-sized star. Stars known as giants may be eight to 100 times as large as the sun. Supergiants are even larger. They may be up to 300 times as large as the sun. A star at the end of its life can collapse and become very small—only about the size of Earth.

Even though the sun is only a medium-sized star, it is the largest object in our solar system. Scientists have been able to calculate the sun's mass from the speeds of the planets and the shapes of their orbits around the sun. The sun's mass is nearly two nonillion kilograms—you can write that as a two followed by 30 zeros! The sun has almost 100 percent of the mass in the solar system. The sun is huge when compared to Earth. In fact, the sun has more than one million times the volume of Earth. If you think of the sun as a gumball machine, it would take over one million Earth gumballs to fill the sun gumball machine!

The surface temperature of the sun is 5,500 °C!

1. The sun is a supergiant.   T / F

2. The sun has two million times more volume than Earth.   T / F

3. The mass of the sun is known from observing the speed and shape of the planets' orbits.   T / F

3 Read and underline three elements that make up the sun's atmosphere. Then look and draw arrows from the labels to the corresponding parts of the sun's atmosphere.

Characteristics of the Sun

The sun is a fiery ball of hot gases and has no hard surfaces. It gives off enormous amounts of light and heat. The outer part of the sun is about 5,500 °C. The inner core could be as hot as 15,000,000 °C.

The Sun's Atmosphere

Like Earth, the sun has an atmosphere. The innermost layer is called the **photosphere**. It gives off the light energy you see. The layer above the photosphere is the **chromosphere**. The outermost layer is called the **corona**. When scientists look at the sun with special equipment, they see dark spots, called **sunspots**, moving on the face of the sun. Sunspots are part of the photosphere. They may be the size of Earth or larger. They look dark because they are not as hot as the rest of the photosphere. The number of sunspots increases and decreases in cycles of about eleven years.

Flash Lab

Measuring Shadows
Have a partner measure your shadow at different times during the day. Write down what you find. Describe how your shadow changes as the sun moves.

Sunspots are darker, cooler areas on the sun's surface.

4 **Read and write P (prominences), S (solar flares), or B (both).**

Solar Eruptions

Two types of eruptions that take place on the surface of the sun are prominences and solar flares. A **prominence** looks like a ribbon of glowing gases that leaps out of the chromosphere into the corona. Prominences may appear and then disappear in a few days or months.

A **solar flare** is an explosive eruption of waves and particles into space. Solar flares are similar to volcanoes here on Earth. A solar flare causes a bright spot in the chromosphere that may last for minutes or hours. Along with extra-bright light, solar flares also give off other forms of energy. This energy is powerful enough to interrupt radio and satellite communication on Earth.

1. Eruptions that take place on the surface of the sun. ______

2. Identified by the appearance of bright spots in the chromosphere. ______

3. Similar to volcanoes on Earth. ______

4. They can last a few days or months. ______

5. They move from the chromosphere into the corona. ______

5 **Look and circle a solar flare.**

Solar flares give off more light than other parts of the sun. They emit radio waves, visible light, X-rays, plasma, and other radiation.

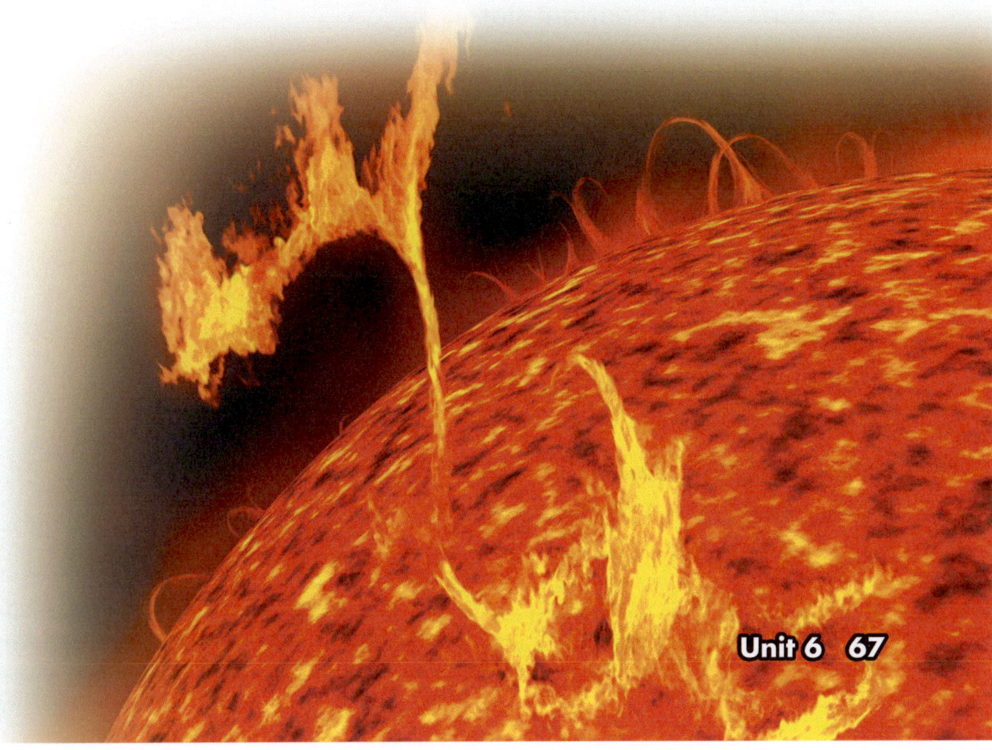

6 **Read and complete the statements. Check answers as a class.**

Constellations

In the past, people looked up at the night sky and "connected the dots" formed by the stars. They saw patterns that reminded them of bears, dogs, and even a sea monster! Today, scientists divide the night sky into 88 constellations. A **constellation** is a group of stars that forms a pattern. Many constellation names are the names of the star patterns that people used long ago.

The star pattern called the Little Dipper contains a star called Polaris. Polaris, or the North Star, is a very hot and very large yellow-white star. It is almost 2,500 times brighter than the sun. It does not look larger than the sun because it is much farther away. Polaris is an important star in navigation. Because it is almost directly above the North Pole, Polaris doesn't seem to move as Earth rotates. If you can find Polaris in the sky, you can tell which direction is north. Early explorers used Polaris as a guide to direct them in their travels. If they located Polaris, then they could determine which direction they were going.

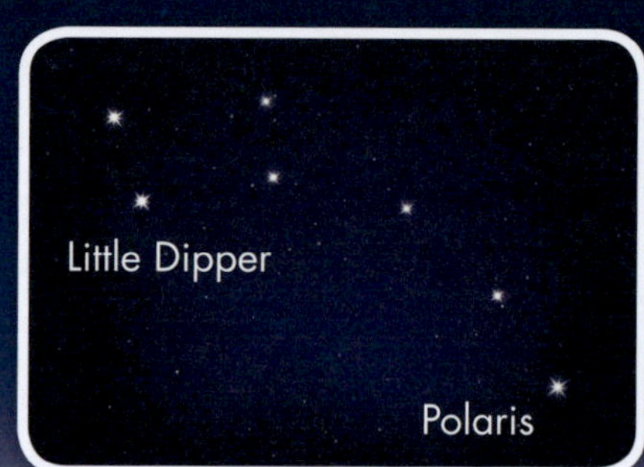

1. Polaris forms part of the star pattern called the ________________.

2. Polaris is almost ________________ times brighter than the sun.

3. Polaris is called the North Star because it is nearly directly above the ________________.

4. Polaris is an important star in ________________.

7 **Read and underline what makes stars appear to move in the sky.**

Stars on the Move

Stars are not always in the same place in the sky. They move in predictable ways. Suppose you looked at the sky early one evening and found the Big Dipper. When you looked two hours later, the Big Dipper seemed to have moved toward the west. Actually, the Big Dipper did not move, but you moved. The spinning of Earth makes the stars appear to move from east to west across the sky.

Think!

How might constellations help scientists study the sky?

 ▷ **Lesson 1 Check** ▷ **Got it?**  **60-Second Video**

Lesson 2 · What are asteroids, meteors, comets, and moons?

1 Look and, with a partner, mark (✓) the rock that might have started out as an asteroid. Explain how you know.

2 Read and underline where the asteroid belt is located. Then, with a partner, discuss how Jupiter is important to Earth.

Asteroids

A rocky mass up to several hundred kilometers wide that revolves around the sun is an **asteroid**. In our solar system, most asteroids orbit in the region between Mars and Jupiter called the asteroid belt. Most asteroids have uneven shapes. Some have smaller asteroids orbiting them. The smallest asteroids are pebble-sized. Most asteroids complete a revolution in three to six years.

Can Earth be hit by asteroids? It has happened, and you can see the huge craters that have resulted. Such collisions are very rare. Fortunately, Jupiter's gravity holds most asteroids in the area beyond Mars.

3 Read, look, and label each asteroid.

Asteroid Ida has a smaller asteroid orbiting it. The smaller asteroid is Dactyl.

____

____

4 Read. With a partner, make a list of a meteor's characteristics in your notebook. Then compare with another pair.

Meteors

Have you ever seen a shooting star? Shooting stars look like bright lines of fast-moving light that form in the night sky. They last a very short time. Shooting stars are not really stars but meteors.

A **meteor** forms when a meteoroid hits Earth's atmosphere. A **meteoroid** is a small piece of rock moving in space. Meteoroids are boulder-sized or smaller. Most are the size of pebbles or grains of sand. When a meteoroid shoots through the air, it heats up quickly. It gets so hot that it glows as a streak of light. Very bright meteors are called fireballs. Most meteors burn up before they hit Earth's surface. If a meteor does not burn up completely, it may fall to Earth. A piece of a meteor that lands on Earth is called a **meteorite**. Most meteorites are quite small. The biggest known meteorite is in Namibia, Africa, and weighs 60 tons.

5 Look and, with a partner, follow the instructions to calculate the size of the meteor in your notebook. Then, measure the circles and draw an *X* on the circle that best represents the probable size of the meteor.

The diameter of this crater may be 24 times larger than the diameter of the meteor that formed it. Measure the crater from point A to point B.

Meteor Crater, in Arizona, was formed by a meteorite impact.

 I Will Know…

6 **Read and fill in the information chart about comets. Check answers as a class.**

Comets

A frozen mass of different types of ice and dust orbiting the sun is a **comet**. Rocky matter may be frozen in the ice. Comets come from areas of the solar system beyond Neptune. Most pass through the solar system in very stretched out and elliptical paths. Several comets a year may travel into the solar system and orbit the sun. You may not see them, though. Only the largest comets can be seen without a telescope.

At certain times each year, meteor showers take place. These occur when Earth passes through the orbit of a comet. A comet heats up and loses dust and rocky matter each time it orbits the sun. These loose pieces remain in the comet's orbit. When these pieces collide with Earth's atmosphere, they become meteors. Discovering a comet is exciting. How can you discover one? Most comets today are found by people who use telescopes to photograph the sky each night. The photos may show a fuzzy object. Another clue is that stars stay in the same relative position to other stars, but comets do not. If an unknown object keeps changing position compared to the stars over a few hours or days, it might be a comet. If you are the first person to discover a comet, it could be named after you.

Comets			
Where They Come From	**How They Travel**	**How Often They Pass through Our Solar System**	**How We Can See Small Ones**

7 In small groups, discuss if you have seen a comet and what it looked like. If you haven't, research what a comet looks like and discuss.

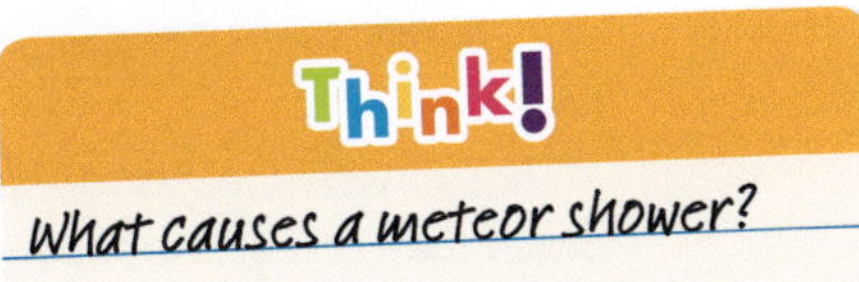

8 **Read and circle the differences between planets and dwarf planets.**

Dwarf Planets

Some objects in the solar system have been classified as dwarf planets. A **dwarf planet** is a large, round object that revolves around the sun but has not cleared the region around its orbit. In 1930, Clyde Tombaugh discovered Pluto. Pluto has an icy solid surface. Astronomers thought for a long time that Pluto was the ninth planet—the only outer planet that is not a gas giant.

Today, astronomers do not consider Pluto a planet. Pluto is a dwarf planet. It is even smaller than Earth's moon. Pluto has an odd orbit. The other planets travel around the sun at the same angle, while Pluto's orbit is tilted. During parts of the orbit, it is closer to the sun than Neptune. This occurred from 1979 to 1999. The next time this will occur is in the year 2227.

Dwarf planets have a similar shape to planets and they revolve around the sun.	Dwarf planets have not cleared the region around their orbit as planets have.	Dwarf planets are much larger in size than planets.	Dwarf planets can also be called moons, unlike planets.

9 **How is Pluto's orbit different from the other planets'? Look and discuss with a partner.**

Charon was the first of Pluto's moons to be discovered. Since then, other moons have been found orbiting Pluto.

Think!

In how many more years will Pluto be closer to the sun than Neptune?

10 **Read and circle *T* (true) or *F* (false). Correct the false statements orally with a partner.**

Moons

The solar system contains many moons. You may remember that a **moon** is a natural object that orbits a body bigger than itself. Like planets, moons often have a spherical shape. Earth and Mars are the only inner planets that have moons. The outer planets have many moons.

The moons in the solar system are very different from one another. Earth's moon has no atmosphere. Saturn's largest moon Titan has an atmosphere so thick that it lets little light pass through. Jupiter's moon Io has volcanoes on its surface that release sulfur. Sulfur gives Io a colorful appearance.

Moon craters like these form when the moon is struck by another object.

1. Moons orbit smaller bodies in the solar system than themselves. **T / F**

2. Mars has several moons because it is an outer planet. **T / F**

3. Earth's moon and Saturn's largest moon are different from one another. **T / F**

4. Jupiter's moon Io has a sulfurous surface. **T / F**

11 **Circle the differences and similarities between moons and asteroids.**

Both asteroids and moons travel in an orbit.

Moons and asteroids cannot collide with Earth.

Asteroids have uneven shapes and can be very small. Moons are larger and often have a spherical shape.

Both moons and asteroids exist in all regions of the solar system.

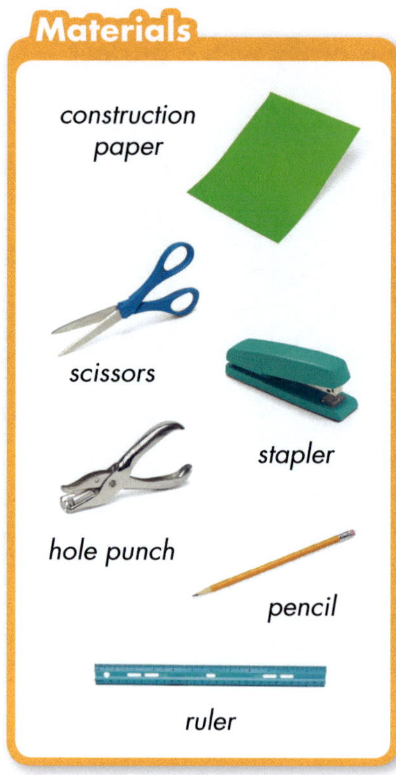

Let's Investigate!

How can spinning affect a planet's shape?

1. Cut 2 strips of construction paper, each 2 cm x 45 cm. Cross them at the center and staple them to make an **X**.
2. Bring the 4 ends together and overlap them. Staple them to form a sphere.
3. Punch a hole through the center of the overlapped ends.
4. Push a blunt pencil through the hole. Only about 5 cm of the pencil should go in.
5. Hold the pencil between your palms. Move your hands back and forth to make your model spin.
6. What shape do you observe when it spins?
7. Record your observations in the table.

Effect of Spinning on a Planet's Shape	
Shape When Not Spinning	**Shape When Spinning**

REVIEW THE BIG ? How do objects move in space?

Lesson 1

What is a star?

1 Circle the answer.
Which layer of the sun gives off the light energy we see?
a) the corona   b) the photosphere   c) the core

2 Fill in the blanks.

> corona   minutes   chromosphere   months   solar flares   eruptions

Prominences and ______________ are ______________ 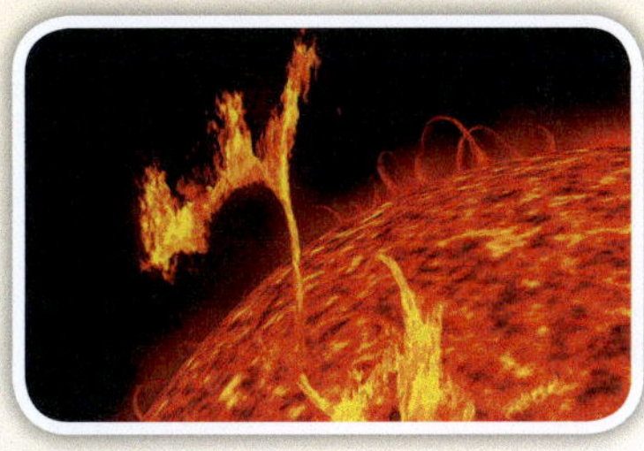
on the surface of the sun. A prominence looks like a
ribbon of glowing gases that leaps into the
______________. Prominences may appear for days
or even ______________. Solar flares appear as bright
spots in the ______________. They may last for ______________ or hours.

Lesson 2

What are asteroids, meteors, comets, and moons?

3 Next to each description, write *asteroids, meteors, comets,* or *moons.*
a) They are rocky and can collide with Earth. ______________
b) They pass through the solar system in elliptical paths. ______________
c) They orbit larger bodies in the solar system. ______________
d) They are small in size and burn in Earth's atmosphere. ______________

4 Look and label the object in the picture.

______________ 

 Got it? Quiz    Got it? Self Assessment

Unit 7

Matter

What are the properties of matter?

I will learn

- some basic qualities of solids, liquids, and gases.
- properties of solutions and how mixtures can be separated.
- that temperature affects many physical and chemical changes.

1 Fill in the blanks.

When soup gets hot enough, some of the liquid becomes _________________ and starts to evaporate. When the soup _____________, you can see bubbles even under its surface.

Ice is water in its solid state. Water becomes ice when it _________________. When it's hot outside, _________________ condenses on the outside of a glass of cold water.

2 Match the photos to the labels.

mixture

solution

3 Label the photos PC (physical change) and CC (chemical change).

Think!

Why does seawater freeze at −2 °C and fresh water at 0 °C?

Lesson 1 · What are solids, liquids, and gases?

1 Read, look at the particles in each picture, and label the state of matter. Then complete the sentences below.

States of Matter

The state of any material is due to the motion and arrangement of its particles. Particles in any state of matter are always moving. Most materials around you are solids, liquids, or gases.

____

Solids A solid has a definite shape and **volume**, or amount of space it takes up. The particles of a solid are very close together. For the most part, they stay in the same place. They do not slide easily past one another. Instead, they vibrate in place.

____

Liquids A liquid has a definite volume but no definite shape. The particles of a liquid can move by gliding past one another. A liquid takes the shape of its container. Forces hold a liquid's particles together, so a liquid keeps a definite volume.

____

Gases A gas does not have a definite volume or shape. The particles of a gas are far apart compared to the particles of solids and liquids. A gas can be squeezed into a smaller volume. If a gas is placed in an empty container, its particles will spread out evenly. The gas will fill all the space and take the shape of that container.

1. For the most part, the particles of a solid ____.

2. The particles of a liquid, by contrast, ____.

3. Compared to solids and liquids, the particles of gases are _______________________________.

2 Read. As a class, discuss why plasma might be a good material for a TV.

Plasmas At very high temperatures, atoms can break down into parts that have electric charges. This state of matter is called **plasma**. Plasma is like a gas because it has no volume or shape of its own. It is also like a metal because it can conduct electricity. The sun is made of gas and plasma.

Explore My Planet!

3 **Read and circle *T* (true) or *F* (false).**

Freezing and Melting

As liquids get colder, their particles slow down. At some point they stop gliding past each other and can only vibrate in place. The liquid becomes a solid. The temperature at which a material changes between liquid and solid states has two names. It is called the **freezing point** when a liquid turns into a solid. It is called the **melting point** when a solid turns into a liquid. Therefore, the melting point and the freezing point are the same temperature.

Each material has its own melting point. Therefore, the melting point can be used to help identify a material. The melting point of lead, for example, is 327 °C.

Some materials are more useful in their solid state than in their liquid state. For example, solid lead is used to weigh down, or sink, fishing hooks.

At its melting point, solid lead becomes liquid and can be poured into molds to give it any shape we want.

1. When liquids get warmer, their particles slow down. **T / F**

2. The melting point of a material is different from its freezing point. **T / F**

3. A material's melting point can identify a material. **T / F**

4. A solid turns into a liquid at its freezing point. **T / F**

4 **What material would be good for a car engine, one with a high melting point or a low one? Discuss with a partner. Then name another thing you think should have a high or a low melting point and explain why.**

Think!

Why do you think the same liquid can be called anti-freeze and coolant?

5 **Read and discuss with a partner how evaporation and boiling are different.**

Evaporation and Boiling

Evaporation takes place when particles leave a liquid and become a gas. Particles evaporate from a liquid when they are at the surface of the liquid and are moving upward with enough speed. This is how rain puddles and the water in wet clothes evaporate.

If the temperature of a liquid is high enough, particles will change to a gas not only at the surface, but also throughout the liquid. As gas particles move quickly upward through a liquid, bubbles of gas form under the surface of the liquid. The **boiling point** of a liquid is the temperature at which this occurs.

You can see the dry salt left behind after the water evaporates from saltwater lakes.

Water boils and evaporates from these hot springs.

6 **Read. Write in your own words what condensation is. Then list another example of condensation not mentioned in the text below and explain how it happens.**

Condensation

Condensation occurs when a gas turns into a liquid. This process often occurs when gas particles touch a cold surface and the temperature of the gas drops. Clouds in the sky and dew on the ground form through condensation of water vapor.

____

____

____

____

____

____

____

At-Home Lab

Wandering Ice

Place an ice cube on a dish and set it where it won't be disturbed. How long does it take for the ice cube to melt? How long does it take for the water to evaporate?

Lesson 2 · What are mixtures and solutions?

1 **Read. Then look at the picture of the fruit mixture and list its separate parts below.**

Mixtures

In a **mixture**, different materials are placed together, but each material in the mixture keeps its own properties. If vegetables are cut and put together to make a mixture, different vegetables do not change their flavors or colors. Most foods that you eat are mixtures of different materials.

Different parts of a mixture can be separated from the rest of the mixture. Suppose your favorite breakfast is a mixture of cereal and raisins. You could easily separate out the raisins with a spoon to eat them first. The parts of a mixture may be combined in different amounts. The bowl of cereal you eat today could have more raisins than the one you ate yesterday.

___________________   ___________________   ___________________

Separating Mixtures

You can use the physical properties of something to separate it from a mixture. The materials in a simple mixture can be separated because they have different physical properties. For example, a magnet can separate iron filings from sand because iron has the property of being attracted by magnets. Sand does not have that property. A screen filter can separate a mixture of pebbles and sand. The smaller particles go through the screen, but the pebbles do not. Sometimes you can sort the parts of a mixture by hand.

2 **What properties do objects in a mixture need to have to be separated by a screen. Write in your notebook.**

3 **With a partner, discuss a mixture you have separated. What was it, and how did you separate its parts?**

4 **Read and list what the solvent(s) and the solute(s) in lemonade are.**

Solutions

A mixture in which substances are spread out evenly and will not settle is called a **solution**. In a solution, the substance that is dissolved is called the **solute**. The substance in which the solute is being dissolved is called the **solvent**. In a solution of sugar and water, the solute is sugar and the solvent is water. Water is sometimes called a "universal solvent" because it can dissolve many substances.

Solutions of a Solid in a Liquid

When a solid dissolves, individual particles separate from the solid and spread evenly throughout the liquid. You can make solids dissolve in a liquid faster by stirring or heating the solution. Grinding a solid into smaller pieces will also help it dissolve faster.

Lemonade is usually made with lemon juice, water, and sugar.

Solvent(s): _________________________   Solute(s): _________________________

5 **Write two ways you can make a solid dissolve in a liquid faster.**

_______________________________   _______________________________

6 **In what ways might water not be a universal solvent? Discuss with a partner.**

Go Green

Did you know that, because water is a good solvent, it's good at cleaning things? That means you can clean things like windows using just water instead of products with chemicals that might harm the environment.

7 Read. Then look at the picture. Write in your notebook what you can tell about the solubility of the objects inside the globe.

Solubility

Many materials can dissolve in water and make a solution. You can dissolve more of some materials than others in the same amount of water. Some materials will not dissolve in water at all. This describes a material's **solubility** in water. Different substances can have different solubility in other solvents.

8 How would putting a jar of water and tea in the sun affect the solubility of tea? What else could you do to affect its solubility? Discuss with a partner.

You can make "sun tea" by putting a jar of water with tea, lemongrass, or hibiscus flowers in the sun.

Flash Lab

Form three small groups. Each group gets a cup of water. One group adds a teaspoon of sugar, a second, a teaspoon of gelatin, and the third, a teaspoon of powdered chalk. Predict whether each substance will dissolve and which will dissolve the fastest or slowest. Then stir each liquid and see whether your predictions are correct.

9 How can you describe a fish tank as both a mixture and a solution? Discuss as a class.

Think!

Why is it important that some gases dissolve in sea water?

Lesson 3 · How does matter change?

1 **Read and underline two examples of how temperature can cause a physical change.**

Physical Changes

Matter changes all the time. Some changes are physical changes. A **physical change** is a change in some properties of matter without forming a different kind of matter. When you cut a piece of paper into smaller pieces, you do not produce a new material. You still have paper. The paper has undergone a physical change. Some of its properties have changed, but the properties that make it paper are still there. For example, the cut pieces are smaller than the original sheet and do not have the same shape. However, these pieces can burn or absorb water. They also keep their original color.

Peeling, slicing, and squeezing fruit into juice are physical changes.

Temperature and Physical Changes

Physical changes can happen more easily or less easily depending on the temperature. For example, a wet towel takes longer to dry in the shade than in the sunlight.

Some physical changes cannot happen unless the temperature is right. For example, ice does not normally melt until its temperature rises above its melting point, 0 °C.

Melting, freezing, evaporation, and condensation are all physical changes. For example, water vapor is still water. We may call it water vapor, but it is just water that has gone through a physical change. In the same way, the melted wax of a candle is still wax. It hasn't turned into a new substance. It has just become liquid. It becomes solid again as soon as it cools off.

2 **As a class, list as many physical changes caused by temperature changes as you can.**

____

____

____

____

____

3 Read and underline three examples of chemical changes.

Chemical Changes

In a **chemical change**, one or more types of matter change into other types of matter with different properties. When a chemical change occurs, atoms rearrange themselves to form new kinds of matter.

It is not always easy to tell if something has changed chemically. Some ways we might know a chemical change occurred include the release of heat and light, a change in color, a new smell, or gas bubbles, for example.

Chemical changes happen all the time around us. The rusting of iron is a familiar chemical change. When you leave an iron object outside, it slowly becomes rusty. Rust is red and brittle. It is a new substance. The process in which plants use water and carbon dioxide to make food is a chemical change because a new substance is made— sugar. When newspapers burn, they also go through a chemical change. They change into ash.

A chemical change inside these bracelets causes them to glow.

A fire leaves behind ash, a new substance.

4 List some examples of chemical changes and some properties that might be affected in a chemical change.

____

____

____

____

____

____

5 Is the process a plant uses to make food a physical or a chemical change? How do you know? Why is this change important for plants? Discuss with a partner.

At-Home Lab

Apples and Limes

Cut an apple into six pieces. Wipe three pieces with water. Put them on a plate labeled *Water*. Wipe the other three pieces with lime juice. Put them on a plate labeled *Lime Juice*. Leave the pieces for 15 minutes. Explain what happened.

6 **Read and underline how high and low temperatures affect chemical changes.**

Temperature and Chemical Changes

When a candle burns, it goes through a chemical change that releases light and heat. But this chemical change cannot start by itself. You need to light the candle. After the candle starts burning, it keeps itself going by the heat it releases.

When a candle burns, the wax and the wick combine with oxygen in the air to become smoke, soot, and hot gas.

Many chemical changes can happen without high temperatures, but they often happen faster if the temperature is high. Remember that particles move faster when the temperature rises, so they may have more chances to rearrange themselves into new substances quickly. For example, if you put a fizzy antacid tablet in a glass of water at room temperature, the bubbles will form faster than if you use cold water.

Low temperatures can slow down chemical changes. When you buy fruit that is not ripe, you can let it ripen in the kitchen at room temperature. Ripening involves chemical changes that slowly change the color and flavor of a fruit. If you put unripe fruit in a refrigerator, the fruit often will ripen more slowly.

7 **Look at the pictures and circle the tomato that was stored in the refrigerator.**

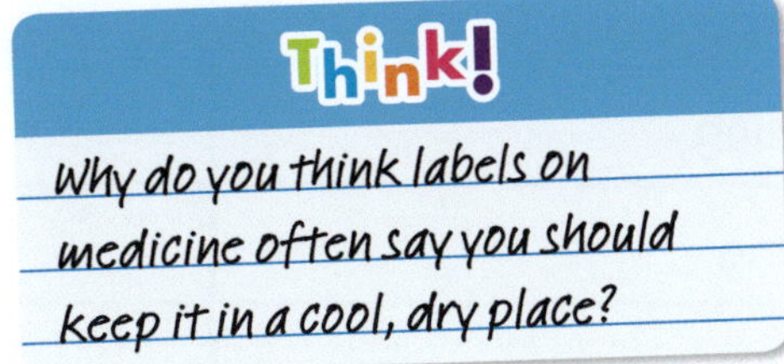

Let's Investigate!

What are some ways to separate a mixture?

1. Put on the safety goggles. Label the 4 cups, *A*, *B*, *C*, and *D*. In Cup A, put 1 spoonful of salt, 2 spoonfuls of sand, 3 marbles, and 100 mL of water.
2. Carefully make 4 holes in Cup B by pushing a pencil through the bottom of the cup from the inside.
3. Hold Cup B over Cup C. All at once, pour the mixture from Cup A into Cup B. Swirl Cup B to clean the marbles. Record the part of the mixture that was removed by straining.
4. Put a coffee filter in Cup D. Slowly pour the mixture from Cup C into Cup D. Record the part of the mixture that was removed by filtering.
5. Remove the filter. Use the spoon to drip 2 drops of the liquid onto the foil. Let it evaporate. Record the results.

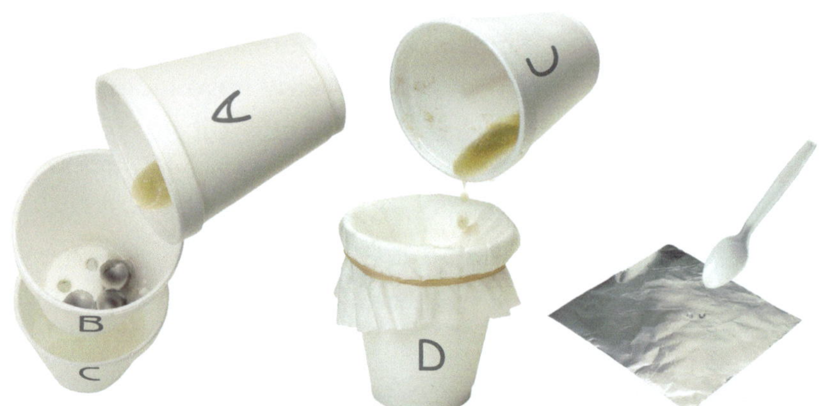

Results of Separation		
Separating Method	**Part Removed**	**Part Not Removed**
Straining		
Filtering		
Evaporation		

What are the properties of matter?

Lesson 1

What are solids, liquids, and gases?

1 **Read and circle *T* (true) or *F* (false). Then correct the false statements.**

a) The temperature at which a solid becomes a liquid is its melting point.   **T / F**
b) The boiling point of water is 327 °C.   **T / F**
c) Gas particles are far apart compared to those of solids and liquids.   **T / F**
d) Liquids only evaporate at their boiling point.   **T / F**

Lesson 2

What are mixtures and solutions?

2 **Read and fill in the blanks.**

Mineral water is a _________________ in which water is the

_________________ and gases are the _________________.

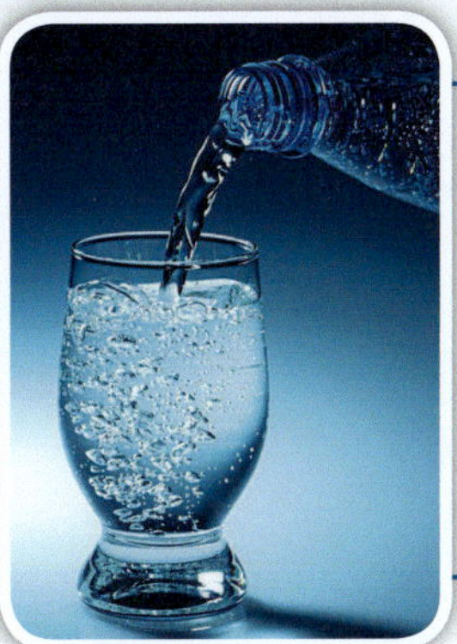

Lesson 3

How does matter change?

3 **Read and label *PC* (physical change), *CC* (chemical change), or *B* (both).**

a) melting, freezing, evaporation, and condensation _______

b) burning a candle _______

c) an avocado ripening _______

4 **Write how physical changes and chemical changes are similar and different:**

Similar ____

Different ____

Unit 8 — Forces and Motion

What affects the motion of objects?

1 Label the photos with the kinds of forces.

I will learn

- about some forces that cause objects to move.
- about different kinds of machines and how they work.

____

____

2 Look at the photo of the golfer. With a partner, identify three forces that will affect the ball.

3 With the class, list the six kinds of simple machines and name some examples of each.

Think!

Why do many stores put up caution signs when they are mopping the floors?

Lesson 1 · What are forces?

1 Read. Then draw arrows to show the direction of the forces in the picture.

Forces

When one object pushes or pulls another object, the first object is exerting a force on the second one. A **force** is a push or pull that acts on an object. Forces can change the way objects move. When an object begins to move, it is because a force has acted on it. When an object is already moving, forces can make it speed up, slow down, or change direction.

Every force has a strength, or magnitude. A force also has a direction. The direction of a force can be described by telling which way the force is acting. If two forces with the same magnitude act in opposite directions, they are called **balanced forces**. The object they act on will not move.

Key Words

- force
- contact force
- balanced forces
- friction
- noncontact force
- air resistance
- gravity
- electromagnets

2 What are three ways you could apply force to act on a merry-go-round once it is already moving?

________________________ ________________________ ________________________

3 With a partner, name as many everyday objects that you push or pull as you can.

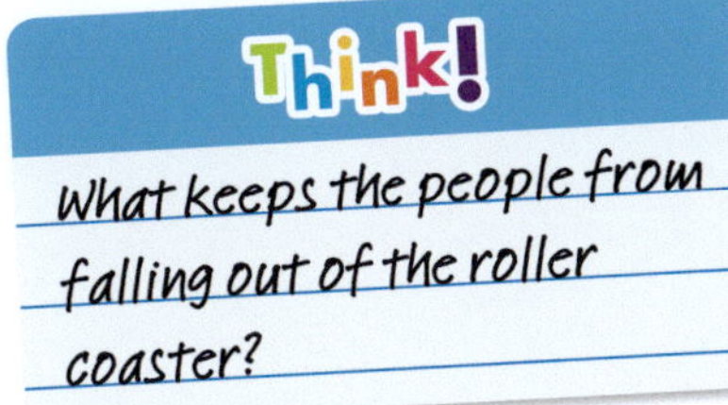

4 **Read. Underline the definition of friction. Then circle five things the amount of friction depends on.**

Contact Forces

Car mechanics use forces to lift tires and pull tool carts. These forces cannot act unless the mechanic touches the object to be moved. The object may be touched directly with a hand or using a handle or a rope, but there must be contact. A force that requires two pieces of matter to touch is called a **contact force**. You exert a contact force when you push or pull a piece of furniture.

One kind of contact force is friction. **Friction** is the force that results when two materials rub against each other or when their contact prevents sliding. Friction makes it harder for one surface to move past another. The amount of friction between two objects may depend on their texture, shape, speed, and weight. It may also depend on whether or not the surfaces are wet.

Solids are not the only materials that can cause friction. Air and water also resist motion when an object pushes against them. **Air resistance** is a type of friction that is present when particles of air contact a surface. Water causes a similar type of friction. Submarines' and ships' shapes are designed to help them reduce friction and move through water easily.

5 **List three examples of sports equipment you have or would like to have. Explain how each is designed to use or reduce friction. Then discuss with a partner.**

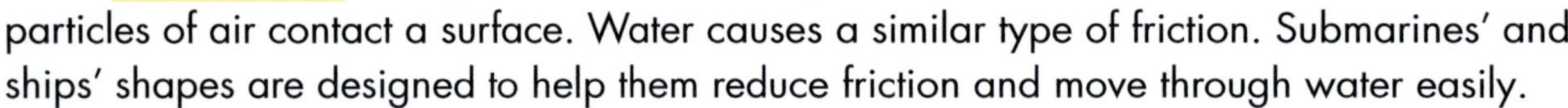

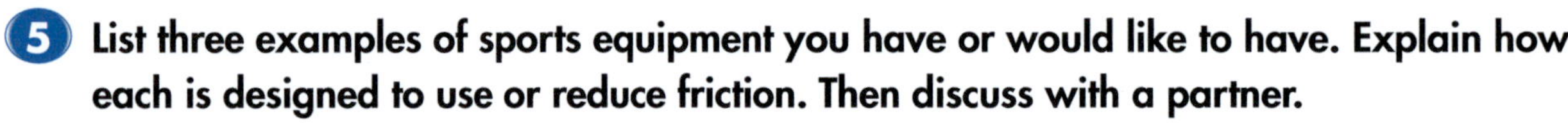

6 **In what ways do you think the design of an airplane takes into account air resistance? Discuss with the class.**

Flash Lab

Air Bags
In small groups, take turns swishing an open, medium-sized plastic bag by its handles so that air fills it up as you swish it. Describe what you can feel.

7 **Read and underline three kinds of noncontact force. Then answer _T_ (true) or _F_ (false).**

Noncontact Forces

For friction to work, two things need to touch. There has to be contact between two surfaces or contact with a gas or a liquid. But there are forces that can act at a distance. They work even if the object that is pushing or pulling is not touching the object being pushed or pulled! A force that acts at a distance is called a **noncontact force**. Three examples of noncontact forces are gravity, electric forces, and magnetic forces.

Gravity

Every object in the universe exerts a pull on every other object. The force of attraction between any two objects is called **gravity**. Only the gravity of a large object such as Earth is strong enough to cause effects that we can notice easily. Without gravity, things would not fall. Gravity pulls objects toward Earth's center without touching them. The weight of an object is the force of Earth's pull on that object. As an object moves away from Earth, the object weighs less and less because the pull of Earth's gravity becomes weaker and weaker with distance.

1. Friction can work at a distance.   **T / F**

2. All objects in the universe have gravitational force.   **T / F**

3 The pull of Earth's gravity affects the weight of objects.   **T / F**

4. Magnetic forces work from a distance.   **T / F**

8 **List some forces that might affect this biker.**

____

____

____

9 **If Earth is pulling down on you, are you pulling up on Earth? Discuss with the class.**

10 **With a partner, research how much you would weigh at sea level and at 300 km above Earth.**

Does Gravity Affect You?
Stand. Stretch your left arm overhead. Leave your right arm at your side. Wait for 1 minute. Then compare the color of the palms of your hands. Share what you notice.

11 **Read. Discuss with a partner why clothing in a clothes dryer might have static electricity.**

Electric Forces

Electric forces act between objects that are electrically charged. Oppositely charged objects are attracted to one another and tend to move toward one another. Objects with the same charge repel each other and tend to move away from one another.

Maybe you have taken clothes out of a clothes dryer and found some pieces sticking together. When you pulled them apart, you may have seen electrical sparks. This is an example of electric force. Because the electrical charges of the two pieces of clothing are not moving, scientists refer to this force as static electricity.

12 **Have you ever had a shock from static electricity? Discuss what it was like and the circumstances in which it occurred with the class.**

13 **Read. Then answer *T* (true) or *F* (false).**

Magnetic Forces

Magnets will pull strongly on objects made of some metals, such as iron, cobalt, and nickel. Every magnet has a north pole and a south pole. Magnetic force is greatest at a magnet's poles. The north pole of one magnet will pull on, or attract, the south pole of another magnet. The north poles of two magnets will push away from, or repel, each other. The south poles of two magnets will act in the same way.

1. Both electric and magnetic forces attract opposites.　　　　**T / F**

2. Two north poles of a magnet repel one another.　　　　**T / F**

3. Magnetic force is greatest at a magnet's center.　　　　**T / F**

With an adult, research on the Internet how to make an electromagnet and then build one.

14 Read. Discuss with the class how maglev trains work and what their advantages might be.

Electromagnetic Forces

Sometimes scientists refer to electric and magnetic forces as a single force called electromagnetic force. They do so because the two forces are related. When electricity flows through a wire, it creates a magnetic force. Engineers use electromagnetic force to build powerful magnets called **electromagnets** that can be turned on or off or that can switch the direction of their north and south poles.

Electromagnets are used in a number of machines, including machines that move heavy metal objects, computer hard disks, and magnetic resonance imaging (MRI).

15 In small groups, research some other electromagnets and what they are used for. Present them to the class.

Why does the electric charge in the air change according to the weather?

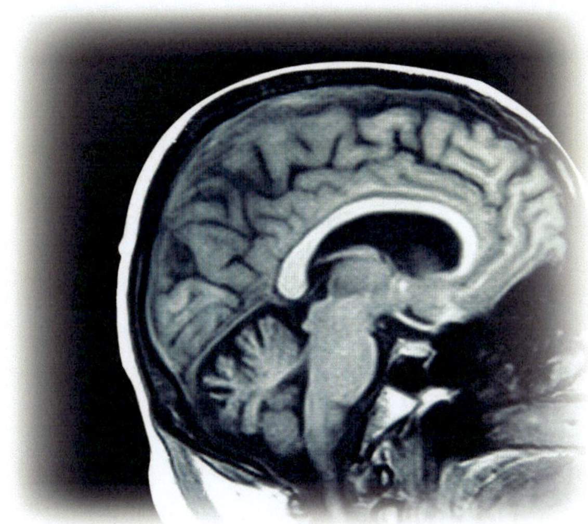

MRI machines use electromagnetic forces to make images of the inside of the body.

Maglev trains use magnets to float over their tracks and to overcome friction.

Lesson 2 · What are machines?

1 Read and underline three ways a simple machine can change your input force. Then answer *T* (true) or *F* (false).

Simple Machines

Work is the energy used when a force moves an object. A machine is a device that helps you do work. A **simple machine** is a machine made up of one or two parts. Machines can help you do work by changing the amount of force required. They may also change the direction of the force or the distance it's applied. Machines do not reduce the amount of work.

When you do work, the force that you use is called the **input force**. A simple machine can increase or decrease your input force, change its direction, or cause it to move an object a longer or shorter distance. The force that the machine supplies is called the **output force**. A nutcracker is a simple machine. The nutcracker turns your small input force into a much larger output force that moves—or cracks—the nut.

There are six kinds of simple machines: the lever, the pulley, the wheel and axle, the inclined plane, the wedge, and the screw.

1. Simple machines are made of many parts.   T / F

2. Machines reduce the amount of work.   T / F

3. The force that the machine supplies is called output force.   T / F

4. Machines can change the distance or direction of force.   T / F

2 Look at the photo and label *input force* and *output force*.

Let's Explore! Lab

3 Read. Suppose you and one of your parents are on a seesaw. What will your parent have to do to make the loads on each end more balanced? Discuss with a partner.

Levers

A nutcracker is a lever. A lever is a type of simple machine in which a bar moves around a fixed point or support called a **fulcrum**. Levers do work by using a bar, the fulcrum, and a force you apply to move a load. A lever like the nutcracker adds to your input force. The input force that you apply to crack a nut is also called **effort**. The object you want to move is called the **load**. Levers can also change the direction of the force.

A seesaw is a lever that changes the direction of a force. When a person pushes down on one end of a seesaw, what's on the other end, the load, goes up. The input force, or effort, occurs when a person pushes down on the seesaw. If the position of the fulcrum changes, the amount of input force needed to move the load changes, too. The farther the fulcrum is from the person using the lever, the easier the load is to lift, but the person has to move the lever a greater distance.

4 Look at the picture of the seesaw. What is wrong with the forces in this photo? What is it trying to communicate? Discuss with the class.

5 How is the effect of a simple pulley different from that of a block and tackle? Think of a use you could have for a block and tackle. Write it in your notebook.

Pulleys

A pulley is a simple machine consisting of a rope or cable that runs around a grooved wheel. A simple pulley changes the direction of the force needed to do work. It does not change the amount of force needed. To lift a load, you must pull on the rope with a force equal to the force with which the load pulls down on the rope. However, a system of two or more pulleys, a block and tackle, reduces the amount of input force needed. A block and tackle has multiple lengths of rope. Each length of rope carries part of the weight of the load. Lifting the load thus requires less effort, but you must pull more rope to do the work.

6 **Read. Label the first photo to show the load, distance, and input force and draw an arrow to show the direction of the input force. Check answers with a partner.**

Inclined Planes

Have you ever tried carrying a heavy load up a flight of stairs? It is much easier to roll it up a ramp! A ramp is an inclined plane. An inclined plane is a simple machine that consists of a flat surface with one end higher than the other.

The input force on an inclined plane is the force that you use to push or pull an object along the plane. An inclined plane decreases the input force required to lift an object, but it increases the distance over which the force is applied. Look at the picture to the right. The distance from the ground to the back of the truck is less than the distance along the ramp. But it takes more force to lift the box from the ground to the truck than to push the box up the ramp. Because the distance increases, the input force is spread over a larger distance when a ramp is used.

Several factors are related to the amount of force needed to move an object up an inclined plane. One factor is the steepness of the plane. Suppose you have two ramps that are the same height but different lengths. The shorter ramp will have a steeper incline. It will take more force to push an object up the shorter, steeper ramp than it will to push the same object up the longer ramp. Another factor is the weight of the object being moved. It takes more force to move a heavier object than it takes to move a lighter object.

7 **With the class, discuss how a ramp can help a person in a wheelchair and what factors might change the amount of force needed to move a wheelchair up a ramp.**

8 Read. Then mark the wedges (✓) and mark the screws (✗).

Wedges and Screws

A wedge is a simple machine that can be made of one or two inclined planes. Some wedges, such as some axes, are made of two inclined planes placed back to back.

A screw is a simple machine consisting of a smooth cylinder with a tiny inclined plane wrapped around it. Screws can be used to pull two pieces of wood together or used in jacks that lift cars. When you use a screwdriver, you provide an input force that turns the screw. The threads of the screw, or the inclined plane, increase the distance the screw travels. This reduces the input force needed to turn the screw.

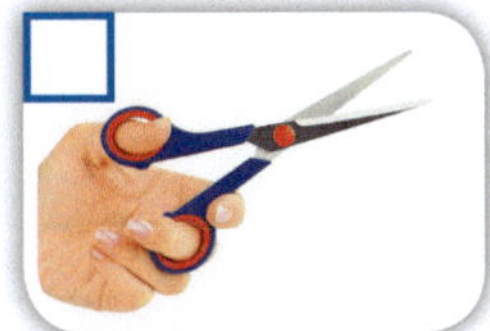
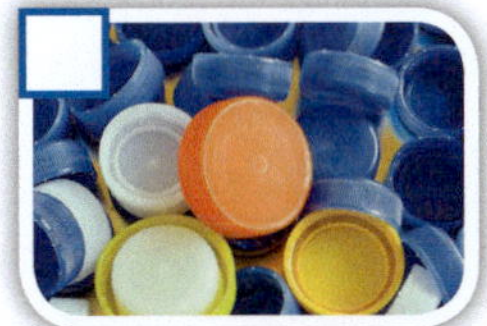

9 Read. Then name two more examples of a wheel and axle.

Wheel and Axle

A wheel and axle is a simple machine made up of a circular object attached to a bar. You make the bar turn by turning the circular object. A wheel and axle reduces the amount of force needed to do work. How hard is it to tighten a screw with your fingers? After the first few turns, the amount of force required is too great. You need a screwdriver, a type of wheel and axle. The handle is the wheel. The metal rod that turns the screw is the axle. It takes less force to turn a screw with a screwdriver.

10 Read. Then label the simple machines that are part of the bicycle.

Complex Machines

A machine that uses two or more simple machines is called a complex machine. Many complex machines use electricity, gravity, burning fuel, human force, or magnetism to operate each of the simple machines within them.

A bicycle is a complex machine.

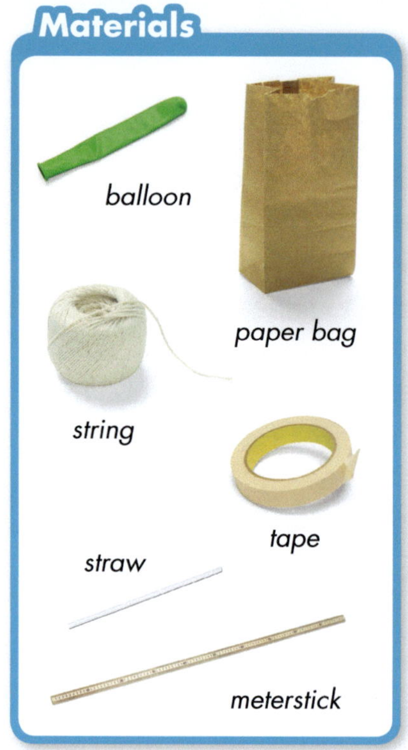

Let's Investigate!

What forces affect the motion of a rocket?

1. Tie one end of a 10-meter piece of string to a chair. Slide a straw onto the string. Tape a paper bag to the straw. Tie the other end of the string to another chair. Make the string tight by pulling the chairs apart. Slide the bag to the middle of the string.
2. Blow up a long balloon. Hold the neck end closed. Put the other end in the bag.

Do not blow up the balloon too much.

3. Let go of the balloon. What happened?
4. Slide the bag to one end of the string. Blow up the balloon again. Place the balloon in the bag. Predict how far the rocket will move. _______________________________________
5. Let go of the balloon. Use a meterstick to measure how far the rocket moved. Repeat 2 more times.
6. Record your data in the table below. Find the average distance for the 3 trials. (Add the 3 distances and divide by 3.)

Rocket Data	
	Distance (m)
Trial 1	
Trial 2	
Trial 3	
Average	

7. Compare how far the rocket moved in each trial. What caused the rocket to move in the direction it did?

____
____
____

8. What made the rocket move? How do you know?

____
____
____
____

What affects the motion of objects?

Lesson 1

What are forces?

1 Mark the following statements *T* (true) or *F* (false).

Friction is a type of noncontact force.	T / F
Air resistance is a type of friction.	T / F
If you push a door harder, you increase its output force.	T / F

2 Explain how the zigzag design of this street makes it easier for cars to go up.

____

____

Lesson 2

What are machines?

3 Look at the wheelbarrow. Name what kind of simple machine it is, circle the input force, and mark (*x*) the load.

____

4 Match the simple machines to the labels.

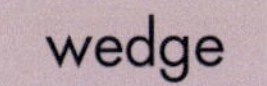

wheel and axle	pulley	wedge

▷ **Got it? Quiz**   ▷ **Got it? Self Assessment**

Energy

How is energy transferred and transformed?

I will learn

- what potential and kinetic energy are.
- the different forms into which energy can change.
- how to describe sound energy.
- how to describe light energy.

1 With a partner, list as many types of energy as you can. Write them in your notebook.

2 Identify the type or types of energy exemplified in these pictures. Discuss as a class.

3 How are sound and light energy similar and different? Discuss as a class.

Think!

If energy cannot be destroyed, why do we say the world is running out of energy?

Lesson 1 · What is energy?

1 Read and underline what energy can change. Then match the sentence parts.

Energy

In science, **energy** is the ability to do work or cause a change. Energy can change an object's motion, color, shape, temperature, or other characteristics. Energy cannot be made or destroyed, but it can change form and can move from one object to another. For example, the energy that a windup toy has when it begins to move does not disappear when the toy stops moving. The energy simply changes into other forms.

A jumping penguin has energy. Some of the energy is due to the penguin's motion, and some is due to its high position during the jump. The sum of the energy due to the motion of an object and the energy stored in the object due to the object's position is called mechanical energy.

Energy cannot be made or destroyed,	does not disappear when it stops moving.
Energy is the	but it can change form and can move from one object to another.
The energy a windup toy has when it begins to move	ability to do work or cause a change.
Mechanical energy is	the sum of energy due to the motion of an object and its stored energy due to its position.

2 What happens to a windup toy's energy when it stops moving? Discuss with a partner.

3 Circle which penguin in the photo above has the most mechanical energy. Discuss with the class how you know.

4 **Read and underline the definition of gravitational potential energy and circle the definition of elastic potential energy.**

Potential Energy

An object does not need to be moving to have energy. **Potential energy** is stored energy. It is not causing any changes now, but could cause changes in the future.

Gravitational Potential Energy

One kind of potential energy depends on the position of an object relative to Earth. This is called gravitational potential energy. For example, as a pole vaulter reaches the top of her jump, she gains potential energy relative to the ground. The higher she goes, the more potential energy she has. She loses potential energy as she falls. The potential energy does not disappear. It just changes form into kinetic energy.

Elastic Potential Energy

Elastic potential energy is the energy of a stretched rubber band or a compressed spring. This type of potential energy is present when things are bent or stretched. For example, when the pole vaulter's pole bends, elastic potential energy increases. The more a rubber band is stretched, the more elastic potential energy it has.

5 **Look at the photos and label *EP* (elastic potential energy), *GP* (gravitational potential), or *B* (both). Discuss your answers in small groups.**

Flash Lab

Rubber Band Release
Wear safety goggles. Stretch a rubber band as if to shoot it, but do not let it go. The energy in the stretched rubber band is potential energy. Aim the rubber band at a blank wall and let it go. How did the energy of the rubber band change?

 I Will Know...

6 Read. Would you have more kinetic energy if you were running or riding your bike?
Write how you know in your notebook.

Kinetic Energy

Kinetic energy is the energy due to motion. The amount of kinetic energy in a moving
object depends on its speed and its mass. The faster an object moves, the more kinetic energy
it has. When a carpenter swings a hammer slowly, the hammer has a small amount of kinetic
energy. When a carpenter swings a hammer quickly, it has more kinetic energy and can
push a nail farther.

Similarly, the more mass a moving object has, the more kinetic energy it has. A moving
beach ball has kinetic energy. If it hits a sand castle, it might knock down a wall. A
basketball has more mass than a beach ball. If a basketball, rolling at the same speed,
hits the sand castle, it might flatten the whole castle!

Potential energy can change into kinetic energy. As the pole vaulter falls, her potential
energy becomes kinetic energy. She moves faster and faster.

Kinetic energy and potential energy can be present in the same object at the same time.
For example, an airplane has potential energy because of its altitude and kinetic energy
because it is moving forward.

7 Read. Then discuss how the bracelets and the firefly are alike with a partner.

Energy Can Change Forms

Electrical energy can be changed into many other
forms of energy. Consider an electrical device, such as a
photocopier. The electrical energy changes into light energy
and mechanical energy. The photocopier also changes
electrical energy into sound energy when it beeps.

A lightning bug uses light to attract its mates or prey. The
lightning bug has chemicals in its abdomen that have stored
potential energy. The chemical potential energy changes
into light energy.

*The windmill changes wind
energy into electrical energy.*

8 How many kinds of energy
does a car use or have?
Discuss as a class.

Lesson 2 · What is sound energy?

1 **Read. Number in order the steps that describe how sound energy is transferred.**

Sound and Energy Transfer

Sound is a wave of vibrations that spreads from its source. For a sound to be heard, energy must first cause an object to vibrate. Vibrating objects transmit, or send off, energy as **sound waves** in air. The energy is transferred through the air as the sound waves move. When the energy of the original vibrations reaches your ears, you hear sound.

☐ Vibrating objects transmit energy as sound waves.

☐ Your brain interprets the energy as sound.

☐ Energy causes an object to vibrate.

☐ Some of the energy reaches your ear.

- sound
- sound wave
- vacuum
- frequency
- vocalization
- echolocation

Your vocal cords vibrate when you talk or sing. These vibrations travel through the air as sound waves. The sound waves travel in all directions.

2 **How are your vocal cords and the strings on a guitar alike and different? How could you stop the sound after plucking the strings of a guitar? Discuss with a partner.**

Flash Lab

The String Phone

Work with a partner. Punch a small hole in the bottom of each of two cups. Cut a piece of string about six meters long. Thread it through the cups. So the string cannot come out, knot it inside each cup. Hold the cups so the string is taut. Talk into one cup while a partner holds the other cup over one ear. Take turns talking and listening.

3 Underline what sound can travel through. Then write in your notebook why sound cannot travel through a vacuum.

How Sound Behaves

Sound can travel through solids, liquids, and gases. Sound cannot travel through a **vacuum**, which is empty space with no particles. Without vibrating particles, sound cannot exist. When a sound wave reaches a border between different materials, it might bounce back to make an echo. Or, the sound wave might be absorbed or pass into the second material.

Sound waves travel through different materials at different speeds. In ocean water, sound travels at about 1,500 meters per second. In air at 0 °C, sound travels at about 330 meters per second.

4 If you wanted to soundproof your bedroom, what kind of material would you use? Discuss with a partner.

5 Read. Then research on the Internet and illustrate what a high-pitched sound wave looks like.

Pitch

Compressions are the areas of sound waves where particles are very close together. The number of compressions that pass by a point each second is the wave's frequency. **Frequency** is a measure of how often particles are vibrating. The greater the frequency, the higher the pitch of the sound will be. Smaller animals and objects often make sounds with higher pitches than large animals and objects.

6 Underline two reasons the volume of a sound might be louder. Then write two things you could do to decrease the volume you hear.

Volume

Why are some sounds louder than others? The sound waves of a louder sound have more energy when they get to your ear. This could be because you are close to the source of the sound. It could also be because the source of the louder sound is vibrating more. When the higher energy gets to your ear, your eardrum will vibrate more than if the sound were softer.

7 Label the photos *high-pitched* or *low-pitched*. Check your answers with a partner. Then think of as many loud and soft sounds as you can.

8 Which of the items in the photos above makes the lowest sound? How do you know? Discuss with the class.

9 Read. Then, in small groups, pick an animal from the text and research how it uses sound. Present your findings to the class.

Animals and Sound

Many animals use sound in different ways. Some animals use sound for communication. Some animals use it to help them navigate or to locate objects. For example, dolphins and humpback whales use **vocalization** to communicate over long distances. Bats and dolphins use **echolocation** to locate and identify objects. That means they send out calls and listen to and interpret the echoes that return. Humans can learn to use echolocation, too!

This wolf is communicating with other wolves.

Bats use echolocation to navigate and hunt.

Think!

Why do sound waves travel faster when it's hot outside?

   ▷ **Lesson 2** Check   ▷ **Got it?**  **60-Second Video**

Lesson 3 · What is light energy?

1 Read. Match the photos to their descriptions and write one transparent, one translucent, and one opaque object you've used today.

Light

Light is everywhere. We get it from the sun, lamps, and cell phone screens. It can travel in straight lines, reflect off objects, and bend as it passes around or through objects. There are several kinds of light. The kind we can see is called visible light. You may know that light can pass through some materials and not others. **Transparent** materials let nearly all light pass through. **Translucent** materials let some light pass through, but not all. A material is **opaque** if it does not let any light pass through it. Light hitting an opaque object is either reflected or absorbed.

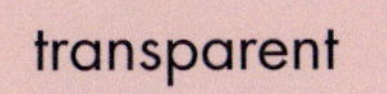

| transparent | translucent | opaque |

2 Write why you think most lampshades are made of translucent materials.

3 Read. Then, with a partner, name one similarity and one difference between visible light and sound.

Light Waves and Color

Like sound, light travels in waves that have certain wavelengths. Also like sound, the speed of light is different in different materials. However, light is different from sound in many ways. For example, light can travel through empty space.

For you to see an object, it must give off or reflect waves of visible light, which must enter your eyes. Different colors have different wavelengths. When all the colors of visible light are mixed, the result is white light. If you shine white light on an object, some wavelengths bounce off it and some are absorbed. The wavelengths that bounce off can enter your eyes. They determine the color you see.

> **Let's Explore! Lab**

4 Read. Then, with a partner, research the electromagnetic spectrum on the Internet and indicate X-rays, visible light, microwaves, and radio waves on the diagram below.

Electromagnetic Spectrum

Unlike sound, light is not a vibration of particles. Light is an electromagnetic wave—that is, a combination of electrical and magnetic energy. Electromagnetic waves can have very long or very short wavelengths. The full range of electromagnetic wavelengths is called the electromagnetic **spectrum**. It includes visible light, but also short wavelengths, such as those of X-rays, and longer wavelengths, such as those of radio waves.

Heat

When you feel the warmth of the sun or the heat of a light bulb shining a few centimeters from your hand, you are feeling electromagnetic waves. Infrared waves feel especially warm. Warm objects produce more infrared waves than cold objects.

Shining Through

In a room that can be made very dark, turn off the lights and turn on a flashlight. Cover the bulb end with your hand. Does some light still shine through? Try blocking the light with other objects. Report what happens.

An infrared photograph shows that some parts of this house are hotter than others.

When food absorbs microwave radiation, its temperature rises.

Radio waves are used for radio, television, and astronomy.

5 What type of light energy does sunscreen protect against? ___________________________

6 Read. Label each picture *reflection*, *refraction*, or *dispersion*. Check your answers with the class.

Light Changes Direction

Light moves in straight lines. When light hits an object, some light is reflected. **Reflection** happens when light bounces off an object. The light still moves in a straight line but goes in a different direction. Mirrors reflect light very well.

Light bends whenever it enters a new material at an angle different from 90°. For example, light bends when it goes from water to air. This bending is called **refraction**. A lens is a polished piece of glass that makes things look larger or smaller when you look through the lens. Lenses do this by refracting the light. Microscopes, cameras, and prescription glasses have lenses.

Different colors refract at different angles. Since white light is a mixture of colors, these colors separate, or **disperse**, when white light is refracted. Rainbows form because water droplets in the air disperse white light into its colors.

      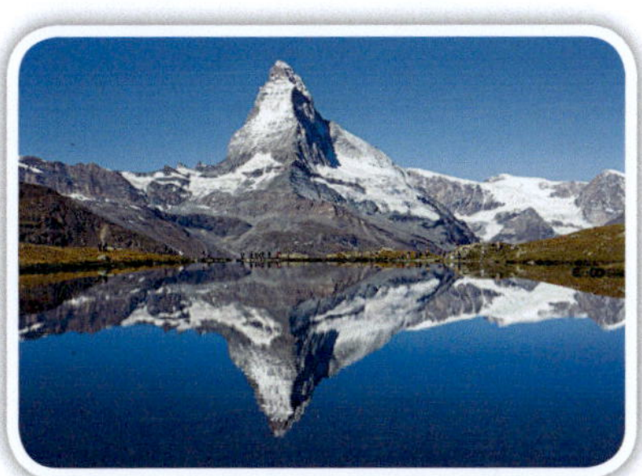

________________   ________________   ________________

7 Look at the picture of the prism. What colors refract more than green, and which refract less? Discuss as a class.

8 How are reflections and echoes similar? Discuss with a partner.

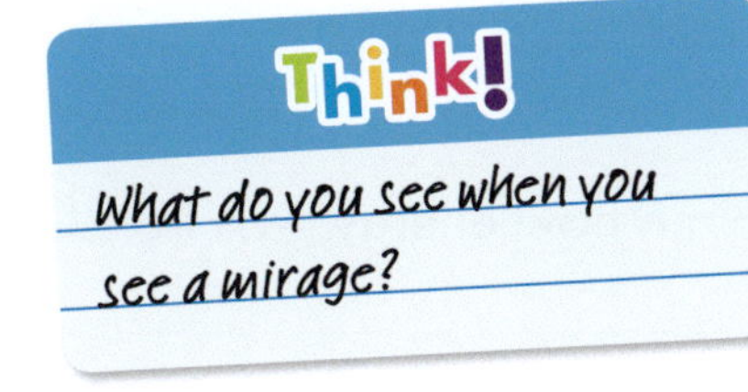

This mirage makes it look like there is water in front of the mountains.

Think!

What do you see when you see a mirage?

Let's Investigate!

How can electrical energy change forms?

1. Make a circuit. Use wires to connect each side of a battery holder to each side of a bulb holder.

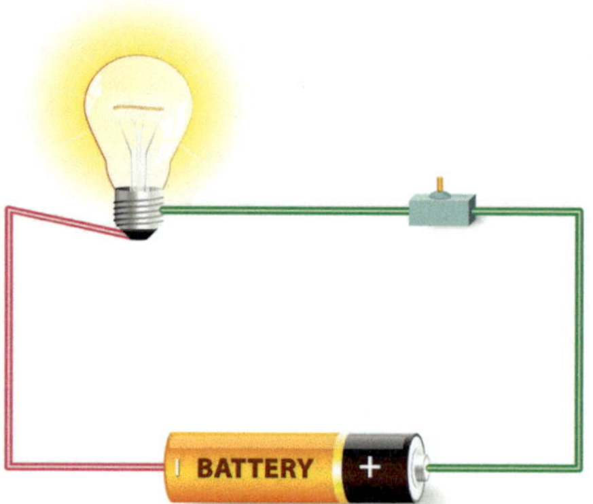

2. In which part of the circuit do you observe light?

____

3. Touch the bulb of the thermometer against the flashlight bulb for 1 minute. What do you observe?

____

4. Make a diagram of your circuit. Show where you observed light energy and thermal energy. Identify the source of the electrical energy.

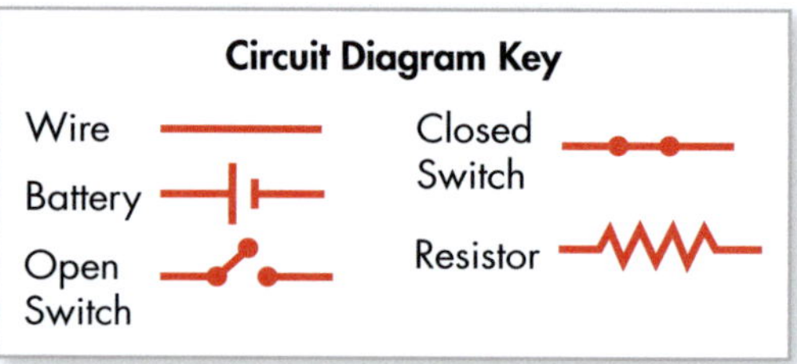

5. Explain how energy was transformed in your investigation.

____

____

____

  ▷ **Let's Investigate! Lab**

REVIEW THE BIG Q

How is energy transferred and transformed?

Lesson 1

What is energy?

1 Fill in the blanks.

A photocopier transforms _____________________ into _____________________ and

_____________________.

Lesson 2

What is sound energy?

2 How does sound travel? Circle *T* (true) or *F* (false).

a) Sound waves travel through all materials at the same speed.　　　　T / F

b) Sound waves travel in all directions.　　　　T / F

c) For sound to be heard, energy must cause an object to vibrate.　　　　T / F

d) The greater the frequency, the higher the pitch.　　　　T / F

Lesson 3

What is light energy?

3 List three components of the electromagnetic spectrum.

_____________________　_____________________　_____________________

4 Write in your notebook what happens when light is refracted, dispersed, and reflected.

Glossary

Unit 1 Design and Function

atom A fundamental building block that makes up all matter.

bristle A short stiff hair, wire, etc. that forms part of a brush.

design process A set of steps for developing products and/or ways to solve problems.

document (v) To record information. Engineers document what they learn throughout the design process to keep track of information and to communicate it to others.

nanobot An extremely small robot that engineers hope to develop.

nanotechnology An area of study that investigates how to develop and make extremely small but powerful machines.

plaque A harmful substance that forms on your teeth and in which bacteria can reproduce.

prosthetic limb An artificial arm, hand, leg, or foot designed to replace a missing one.

prototype A working model that tests a design.

robotic Related to robots.

sensor system A mechanism in robots that mimics the human nervous system, including the brain.

Unit 2   Survival and Extinction

adaptation   Physical and behavioral features that help animals and plants survive.

coal   A hard, black mineral that comes from fossils and is burned to produce heat.

extinct species   A species that was once alive but no longer has any living members.

fossil   An animal or plant that lived thousands of years ago or its shape that has been preserved in rock.

fossil fuel   An energy source, such as coal and oil, that comes from the remains of organisms that lived millions of years ago.

hadrosaur   A dinosaur that had a large, hollow crest on its head.

hibernation   A state in which animals conserve energy by slowing down their body functions.

instinct   A behavioral adaptation that is inherited.

migrate (v)   To travel from one part of the world to another.

oil   A thick, dark liquid that is burned to produce heat and also is made into gasoline.

paleontologist   A scientist who studies fossils.

sauropod   A dinosaur that had a small head, long neck, and enormous body. Sauropods were herbivores that may have used their long necks to reach leaves on tall trees.

Unit 3   Body Systems and Functions

air sacs   Small sacs in the lungs where oxygen enters the blood and carbon dioxide leaves it.

artery   A blood vessel that carries blood away from the heart.

brain   The organ that controls the nervous system.

bronchioles   The smallest branches leading from the trachea to the air sacs in the lungs.

capillary   The smallest blood vessel.

circulatory system   The system that moves blood through the body.

diaphragm   The dome-shaped muscle that moves down when you inhale and relaxes when you exhale.

exhale (v)   To breathe out.

heart   A muscular organ that pumps blood throughout the body.

inhale (v)   To breathe in.

lungs   The organs in which the body exchanges oxygen and carbon dioxide.

nervous system   Receives information from the senses and controls how the body responds.

organ   Different tissues joined into one structure to perform a main function in the body.

respiratory system   The system of the body that helps you breathe.

system   A set of things that work together as a whole.

tissue   A group of the same kind of cells that work together to do the same job.

trachea   The tube that carries air to the lungs from the larynx.

vein   A blood vessel that transports blood toward the heart.

Unit 4   Water and Weather

altitude The height in the atmosphere above sea level.

atmosphere The blanket of air that surrounds Earth.

barometric pressure The pushing force of the atmosphere.

circulation The movement of air that redistributes heat on Earth.

equator The imaginary line that circles Earth midway between the North and South Poles.

hail Precipitation that freezes in layers before falling to Earth.

humidity The amount of water vapor in the air.

jet stream A narrow band of high-speed wind high in the atmosphere. A polar jet stream blows from west to east over North America.

meteorologist A scientist who studies and predicts weather.

sleet Precipitation in the form of frozen raindrops.

snowflake Precipitation in the form of ice crystals that grow larger and start to fall toward Earth.

trade winds The persistent pattern of winds that blow near the equator.

weather The state of the atmosphere, including its temperature, wind speed and direction, air pressure, moisture, amount of precipitation, and other factors.

Unit 5 Earth's Surface

biomass fuel A fuel source made by living things or from recently living things.

constructive force A natural activity that builds new features on Earth's surface.

converging boundary An area where two plates move into each other.

destructive force A natural activity that wears away or tears down features on Earth's surface.

fault A crack in Earth's lithosphere where two plates meet.

inexhaustible resource A resource that will not run out.

landfill A place where garbage is buried under the ground.

lithosphere The solid, rocky outermost layer of Earth.

litter Garbage that people throw onto the ground.

nonrenewable resource A resource that cannot be replaced or cannot be replaced as fast as people use it.

plate A section of the lithosphere.

pollutant An unwanted substance added to the water, air, or soil.

renewable resource Resources that can be replaced.

sliding plate boundary An area where two plates move past each other in opposite directions.

spreading plate boundary An area where two plates move apart from each other.

Unit 6   Earth and Space

asteroid A rocky mass up to several hundred kilometers wide that revolves around the sun.

chromosphere The layer above the photosphere in the sun's atmosphere.

comet A frozen mass of different types of ice and dust that orbits the sun.

constellation A group of stars that forms a pattern.

corona The outermost layer of the sun's atmosphere.

dwarf planet A large, round object that revolves around the sun but has not cleared the region around its orbit.

meteor What forms when a meteoroid hits Earth's atmosphere.

meteorite A piece of a meteor that lands on Earth.

meteoroid A small piece of rock moving in space.

moon A natural object that orbits a body bigger than itself.

photosphere The innermost layer of the sun's atmosphere.

prominence A ribbon of glowing gases that leaps out of the chromosphere into the corona.

solar flare An explosive eruption of waves and particles from the sun's surface into space.

sunspots Areas of the photosphere that appear to be darker because they have a lower temperature. Sunspots may be the size of Earth or larger.

Unit 7   Matter

boiling point The temperature at which a liquid will change to a gas throughout its volume.

chemical change A change that creates new types of matter with different properties.

condensation The process through which a gas turns into a liquid.

evaporation The process through which particles leave the surface of a liquid and become a gas.

freezing point The temperature at which a liquid turns into a solid.

melting point The temperature at which a solid turns into a liquid.

mixture A collection of different materials that are placed together, in which each material keeps its own properties.

physical change A change in some properties that does not create a different kind of matter.

plasma A state of matter that is composed of charged and neutral particles that are not closely associated as atoms.

solubility The capacity of a substance to dissolve.

solute The substance that is dissolved in a solution.

solution A mixture in which substances are spread out evenly and will not settle.

solvent The substance in which a solute is dissolved.

volume The amount of space something takes up.

Unit 8   Forces and Motion

air resistance A type of friction that is present when particles of air contact a surface.

balanced forces Two forces with the same magnitude that act in opposite directions.

contact force A force that requires two pieces of matter to touch.

effort The input force that someone applies.

electromagnets Powerful magnets that use electricity to create magnetic forces.

force A push or pull that acts on an object.

friction The force that results when two materials rub against each other or when their contact prevents them from sliding past each other.

fulcrum A fixed point or support that a lever turns on.

gravity The force of attraction between any two masses.

input force The force you apply to do work.

load The object someone wants to move.

noncontact force A force that acts at a distance. Gravity is a noncontact force.

output force The force that a machine produces.

simple machine A machine made up of one or two parts. An inclined plane is a simple machine.

Unit 9   Energy

disperse (v) To separate. When white light is refracted by a prism, the colors of the visible spectrum disperse.

echolocation An ability some animals have to send out sound waves and locate objects based on the echoes that return from those objects.

energy The ability to do work or cause a change.

frequency A measure of how often particles are vibrating.

kinetic energy Energy due to motion.

opaque Used to describe objects that do not let any light pass through them.

potential energy Stored energy.

reflection What occurs when light bounces off an object.

refraction What occurs when light bends because it enters a new material at an angle different from 90°.

sound A wave of vibrational energy that spreads from its source.

sound wave Energy transmitted by vibrating objects.

spectrum The distribution of a particular form of energy. We can only see light within the visible spectrum.

translucent Used to describe objects that let some light pass through them.

transparent Used to describe objects that let nearly all light pass through them.

vacuum Empty space with no particles.

vocalization The sounds that some animals make in order to communicate.

Printed in France by Amazon
Brétigny-sur-Orge, FR